Illuminating Disease

Illuminating Disease

An Introduction to Green Fluorescent Proteins

Marc Zimmer

Oxford University Press is a department of the University of Oxford.
It furthers the University's objective of excellence in research, scholarship,
and education by publishing worldwide.

Oxford   New York
Auckland   Cape Town   Dar es Salaam   Hong Kong   Karachi
Kuala Lumpur   Madrid   Melbourne   Mexico City   Nairobi
New Delhi   Shanghai   Taipei   Toronto

With offices in
Argentina   Austria   Brazil   Chile   Czech Republic   France   Greece
Guatemala   Hungary   Italy   Japan   Poland   Portugal   Singapore
South Korea   Switzerland   Thailand   Turkey   Ukraine   Vietnam

Oxford is a registered trademark of Oxford University Press
in the UK and certain other countries.

Published in the United States of America by
Oxford University Press
198 Madison Avenue, New York, NY 10016

© Marc Zimmer 2015

All rights reserved. No part of this publication may be reproduced, stored in a
retrieval system, or transmitted, in any form or by any means, without the prior
permission in writing of Oxford University Press, or as expressly permitted by law,
by license, or under terms agreed with the appropriate reproduction rights organization.
Inquiries concerning reproduction outside the scope of the above should be sent to the
Rights Department, Oxford University Press, at the address above.

You must not circulate this work in any other form
and you must impose this same condition on any acquirer.

Library of Congress Cataloging-in-Publication Data
Zimmer, Marc.
Illuminating disease : an introduction to green fluorescent proteins / Marc Zimmer.
  pages cm
Includes bibliographical references and index.
ISBN 978–0–19–936281–3 (alk. paper)
1. Green fluorescent protein.   I. Title.
QP552.G73Z57 2015
572'.645—dc23
2014014222

CONTENTS

Acknowledgments   vii

  1. Nature's Lights   *1*
  2. Heart Disease   *23*
  3. Chagas   *43*
  4. Malaria   *53*
  5. Dengue Fever   *88*
  6. Cancer   *109*
  7. Influenza   *136*
  8. HIV/AIDS   *154*
  9. Diseases of the Brain   *166*
10. Optogenetics   *203*

Index   219

ACKNOWLEDGMENTS

Nature has supplied us with a spectacular light show. Nature's lights are so alluring and intriguing that, as far back as we know, scientists have been trying to unlock their secrets. This is just as well, for who would have imagined that the knowledge gained by understanding the jellyfish's green pinpricks of light, the firefly's come-hither flashes, and the coral's fluorescence would open a Pandora's box of scientific tools? Before anything else, I have to acknowledge all the bioluminescent and biofluorescent organisms on earth and the scientists who have used these organisms' lights to illuminate disease; without them, there would be no book. I particularly want to thank the researchers who spoke to me about their work and shared their images.

At Connecticut College, I have been fortunate in having an inspiring friend and mentor. Although not a scientist, Gene Gallagher has shown me how to make the most of academic life and how to enjoy a life of learning, writing, and teaching. We share a love for music, trashy novels, travel, red wine, and engaged students. On writing days he sends me modified song lyrics about the number of words he has written that day, and I send him links to interesting scientific papers; I get "No Words No Cry," and he gets fluorescent foreskin cells. Strangely enough, it works, and it has made teaching and writing this book lots of fun. It helps that he comes with a great family, too.

Without Bruce Branchini's firefly luciferase research and Doug Prasher's GFP talk, I never would have stumbled across the fluorescent protein family; and what a colorful, versatile, and useful family of proteins they are. The Connecticut College chemistry department and the Connecticut College students have all been supportive and made researching fluorescent proteins a pleasure. John Lusins, Amy Nemser, Andrew Warren, Maria Donnelly, Flavia Fedeles, Ming Chen, Justin Urgitis, Shahzad Zaveer, Lopa Desai, Nana-Yaa Baffour, Scott Maddalo, Curren Mbofana,

Nathan Lemay, Alicia Morgan, Elizabeth Archer, Luisa Dickson, Colleen Megley, Alex Samma, Chelsea Johnson, Shuang Song, Samuel Alvarez, Wayne Ong, Ivan Leroux, Ramza Shahid, Paola Peshkepija, Shawn Mulcahy, and Binsen Li are all Connecticut College undergraduate students who have done research with me and have published GFP-related papers. I have enjoyed working with every one of them. Thanks to their research, I also have to thank the National Institutes of Health for funding my GFP research. Julia McGinly has illustrated a children's book I wrote about GFP. *Jellybelly and the Prettyboys* has the most amazing illustrations I have ever seen. Although we don't have a publisher yet, her artwork is stunning and should be seen by everyone. I will always be a fan.

As a professor at Connecticut College, a liberal arts college that equally values my research, teaching, and outreach, I have the greatest job in the world. I get to interact with passionate and lively students who appreciate my fluorescent axolotls and seem to enjoy my "Glow" freshman seminar. And in Roger Brooks, the dean of the faculty, I have had an interested and supporting colleague who has made it possible for me to take my love for fluorescent proteins on the road to scores of local high schools and middle schools, and even on a trip around the world.

Fluorescent proteins have taken my family and me on a voyage around the Mediterranean and the world. The *MV Explorer* is a modern passenger ship that takes about 600 college students a semester and circumnavigates the globe with them. The students are participants in a global comparative study-away program, and each semester they are taught by a select group of faculty hired from all over the world. Since hearing about the program, I have wanted to teach on the ship and have tried to devise a comparative global chemistry classes to teach. Before becoming a book, *Illuminating Disease* was the class I developed for Semester at Sea. In the spring 2012 semester, 16 students took the class and learned about fluorescent proteins and Chagas disease (in Brazil), malaria (in Ghana), AIDS (with a field trip to an AIDS clinic in South Africa), dengue fever (India and Vietnam), cancer (China), heart disease (with a field trip to the RIKEN Research Center in Japan), and brain disease (United States). I thank Victor Luftig, the academic dean on the voyage, for hiring me and for making the course and this book a possibility.

Jean Koreltiz, author of *Admissions* and *You Should Have Known*, read, edited, and provided invaluable support in writing a children's version of *Illuminating Disease* to be published by Lerner, 2015.

Bob Hoffman, from AntiCancer Inc., Prapti Kafle, Zinya Talukder, Bobby Langan, Matthew and Dianne Zimmer have read and commented on most of the book, and Joe Schroeder, Susan Murosako, Karel Svoboda,

Ashley Hanson, Mohamed Diagne all checked a chapter or two for me. Any mistakes that remain are all mine and have probably been added to the manuscript after they read it.

For five years I had the honor of being the Barbara Zaccheo Kohn '72 Professor of Chemistry, and now I am the Jean C. Tempel '65 Professor of Physical Science. I have to thank Barbara and Jean for their generosity to Connecticut College and myself.

Mackenzie Brady from the Sheedy Literary Agency has been an amazing agent who constructively read my manuscript, promptly replied to my silly questions, and found a home for *Illuminating Disease*.

My family—Caitlin, Matthew, and Dianne—has always been there with bemused support when I couldn't resist expounding on another fascinating fluorescent protein application. They couldn't be more perfect for me even if they had some fluorescent properties themselves.

New articles and teaching materials related to this book can be found at http://illuminatingdisease.conncoll.edu/

Illuminating Disease

CHAPTER 1

Nature's Lights

The remarkable brightly glowing green fluorescent protein, GFP, was first observed in the beautiful jellyfish, *Aequorea victoria* in 1962. Since then, this protein has become one of the most important tools used in contemporary bioscience. With the aid of GFP, researchers have developed ways to watch processes that were previously invisible, such as the development of nerve cells in the brain or how cancer cells spread. Tens of thousands of different proteins reside in a living organism, controlling important chemical processes in minute detail. If this protein machinery malfunctions, illness and disease often follow. That is why it has been imperative for bioscience to map the role of different proteins in the body.

> Professor Gunnar von Heijne, *"Glowing Proteins—A Guiding Star for Biochemistry,"* announcement of the 2008 Nobel Prize in Chemistry

Scientific breakthroughs can come from the most unexpected places. Surprising everyone, including the researchers involved, a topic of little apparent relevance can change the way science is conducted. And so it was that one person's interest in a lowly jellyfish changed the way modern biomedical research is conducted. Thanks to the crystal jellyfish, specifics about disease have been illuminated in a new and wondrous way, allowing us to observe facets of medicine that had never been seen before, just as the invention of the microscope more than 300 years ago opened windows to the world too small to be seen with the naked eye.

The crystal jellyfish, *Aequorea victoria*, which drifts aimlessly in the northern Pacific, has no brain, no anus, and no poisonous stingers. It is an unlikely candidate to ignite a revolution in biotechnology, yet on the periphery of its umbrella it has about 300 photoorgans that give off pinpricks of green light. Proteins derived from these tiny light bulbs have illuminated millions of experiments all around the world and changed our understanding of many diseases.

A. victoria are dimorphic jellyfish (figure 1.1). They alternate between asexual polyps and sexual planktonic medusae, the free-floating gelatinous creatures most people know. In late spring, the medusa forms buds off the bottom of living polyps and starts floating close to the surface of the ocean. The timing is unfortunate, for just when most of the summer tourists want to cool off in the warm ocean waters, the jellyfish medusae make their appearance. Fortunately for those studying the crystal jellyfish, they have no stingers. However, it is not all that simple—anybody interested in these jellyfish lights must travel to the Pacific during the early summer months when *A. victoria* is in the medusa stage of its life cycle and catch some crystal jellies. They are found only in the Pacific Northwest and are not around for long.

Osamu Shimomura, a research scientist at Princeton University and Woods Hole Oceanographic Institute, spent more than 40 years trying to understand the chemistry responsible for the emission of the green light in *A. victoria*, and in the process, he caught more than a million jellyfish. Every summer for more than twenty years, Shimomura and his family would make the 3,000-mile drive from Princeton, New Jersey, to the University of Washington's Friday Harbor laboratory, where they would spend the summer days catching crystal jellyfish from the side of the pier (figure 1.2). They were plentiful. According to Shimomura, "A constant stream of floating jellyfish passed along the side of the lab dock every morning and evening, riding with the current caused by the tide. Sometimes they were extremely abundant, covering the surface of the water." (1, p. 7) For 20 years at Princeton and an additional 20 years at Woods Hole Oceanographic Institute, Shimomura spent his days, and sometimes his nights, unraveling the mysteries of the jellyfish's glow.

Although Shimomura was an expert on bioluminescence, for many years the crystal jellyfish resisted his best efforts to understand its photochemistry. Hundreds of thousands of jellyfish would yield only a few micrograms of bioluminescent proteins, and their chemistry was complicated. However, Shimomura persevered, and *Aequorea*'s mysteries were slowly revealed. There were few other scientists who would have had the determination, the experimental skills, and the ability to meticulously and painstakingly isolate the microscopic amounts of proteins needed to do the experiments that were required to tease out their function. A closer look at Osamu Shimomura's background is quite revealing and illustrates where his perseverance originated.

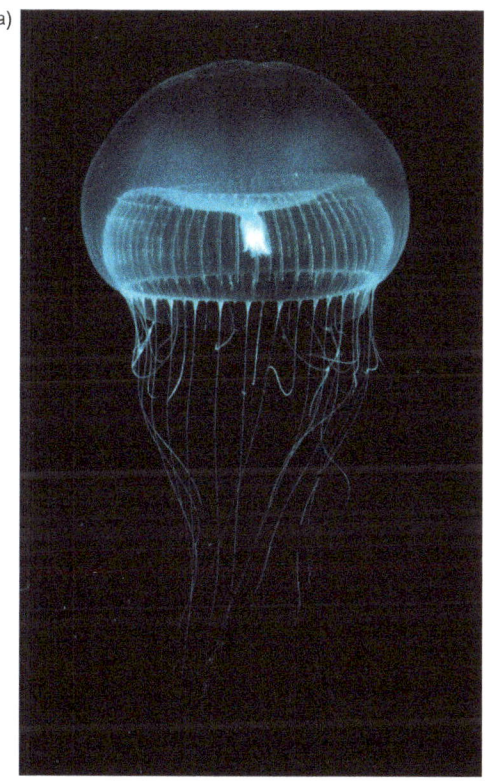

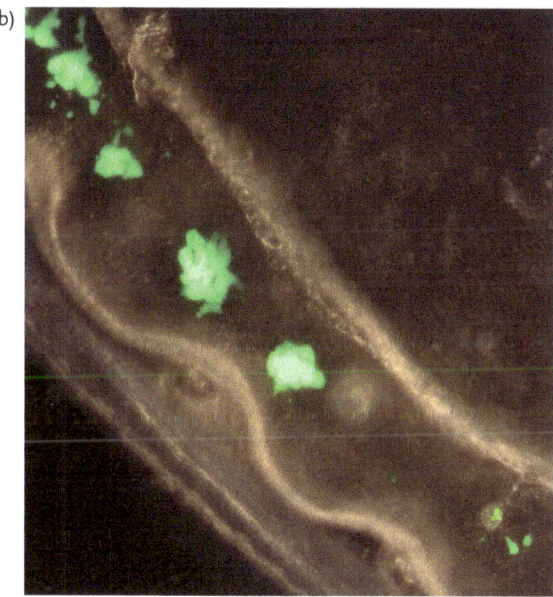

Figure 1.1 The crystal jellyfish, *Aequorea victoria*, medusa (a) has about 300 photoorgans (b) that can emit green light. They are all located on the bottom edge of the jellyfish's umbrella. (Courtesy of Steven Haddock, http://biolum.eemb.ucsb.edu.)

Figure 1.2 The Shimomuras and friends on the pier at Friday Harbor in 1974. Crystal jellyfish were caught with the nets and placed in the buckets. When all the buckets were filled, they were taken to a station wagon parked at the end of the pier and driven to the laboratory, where the circular areas containing the photoorgans were removed from the remainder of the jellyfish and the protein isolation process was started. Osamu Shimomura is standing in the back, third from the right. Thirty to 40 buckets were collected every day. Notice the nets are painted in Princeton colors. (9)

Shimomura was born on August 27, 1928, in Kyoto, Japan. His father was in the Japanese army, and Osamu Shimomura spent his formative years in both China and Japan. When Shimomora was 10, his father was posted to the Soviet border of China, and Osamu was sent back to live with his grandmother in Japan. He recalls, "Grandmother was very strict about manners and etiquette. I always had to keep a good posture in her presence. She often said, 'the samurai betrays no weakness when starving.' After I bathed, she would check behind my ears and neck for dirt. If she found any, she would say it would be ignominious to be dirty when I was beheaded. I knew she was talking about the importance of readiness, but it was a little scary." (2) Shimomura lived with his grandmother for only a year, and fortunately her views on readiness had paled by the time the Second World War reached him.

On the first day of school in 10th grade, Shimomura found out his class was not going to do any schoolwork that year; the students were being conscripted to help with the war effort. Shimomura was assigned to work at the Omura Naval Aircraft Arsenal on the outskirts of Nagasaki, where he repaired fighter engines. The posting continued even after his graduation from high school in March 1945, and on August 9, 1945, when the atomic bomb, named Fat Man, was dropped on Nagasaki, Shimomura was at the factory. He was close enough to see the bomb being dropped and to see the flash associated with the explosion. Sixty-three years later Shimomura recalled that on his walk home, "a drizzling rain started. It was black rain. By the time I arrived home, my white shirt had turned gray. My grandmother quickly readied a bath for me. That bath might have saved me from the ill effects of the strong radiation that presumably existed in the black rain." He is still affected by the memories of the day. (2)

Life in postwar Japan was hard for a 16-year-old boy. After 2 years of looking for a university that was accepting students, Shimomura was extremely fortunate in finding a medical university that was opening a branch in Nagasaki. Shimomura applied for a spot in the pharmacy school and was admitted. There were severe food shortages in Japan, and most students were starving. Osamu was lucky he lived at home and his family owned some farmland, so he didn't have to go hungry. In 1951, he graduated at the top of his class from the Nagasaki Pharmacy School. Jobs were difficult to find in postwar Japan; even with a pharmacy degree, Shimomura struggled before finding a temporary position as a research assistant in Professor Yoshimasa Hirata's laboratory at Nagoya University. The laboratory was interested in establishing the mechanism of cypridina's bioluminescence.

Cypridina, or sea-fireflies, have three eyes, are 3 millimeters long, and live in the sand at the bottom of shallow waters off the southern Japanese coast. During the Second World War, the Japanese army had a unique use for the bioluminescence of cypridina—they were used as miniature flashlights to keep infantry members together while marching through the jungle at night. When crushed, cypridina give off an oily substance, which emits a blue glow. The infantry men would crush the cypridina in their hands and smear the bioluminescent oil on the shirts of their comrades in front of them. The resulting glow allowed them to follow each other without any light being visible to the American forces. After the war, large stocks of unused cypridina remained. Professor Hirata had his fair share of them, as did a much larger laboratory that specialized in bioluminescence studies at Princeton University.

Both labs were interested in determining the mechanism of the cypridian light production. That was a difficult process, since the most important chemicals involved in this process are sensitive to air and decompose when exposed to oxygen. In Hirata's lab it was Shimomura's job to isolate and characterize this air-sensitive compound called *luciferin*. To prevent it from reacting with oxygen, he had to do all the isolations under hydrogen, a very explosive gas. It took about 500 grams of cypridina (a couple of buckets filled to the brim) and five days and nights of constant careful work to purify about 2 milligrams of the cypridian luciferin (just a few grains). Although he had to repeat this tedious procedure many times, Shimomura enjoyed the attention to detail the research required. He did not mind working with minuscule quantities and found the repetitive nature of the research calming. The research into light production in cypridina ignited his lifelong passion for studying bioluminescence. His successes, after all the hard work, felt very rewarding. Shimomura said, "When I finally succeeded, I was so happy I couldn't sleep for three days. Since the end of the war my life had been dark but this gave me hope for my future. Probably the greatest reward I gained was self-confidence; I learned that any difficult problem can be solved by great effort." (9, p5594)

Shimomura's results were remarkable, and his work garnered the attention of the much larger rival group at Princeton, which invited him to come to the United States. Thanks to a Fulbright scholarship, he made the move, and in 1961, he and his wife started what would be a 22-year stay at Princeton University, where he did most of his jellyfish bioluminescence research. (2–6)

Osamu Shimomura found that *A. victoria* uses two proteins to make its green light. (7) The first protein, which he named *aequorin* in honor of *Aequorea*, gives off blue light when it binds calcium in a test tube, but in the photoorgans of the jellyfish, the blue light energy produced by the aequorin is entirely absorbed by a different protein, which subsequently reemits the energy as green light. This process in which high-energy blue light is absorbed and is then immediately returned as lower-energy light green light is called *fluorescence*. That is why the second protein has appropriately been named *green fluorescent protein*, which is usually abbreviated as GFP.

The crystal jellyfish was the first organism known to use two proteins, GFP and aequorin, to make light; there was no precedence in the scientific literature for this type of bioluminescence, and so Shimomura had to break new ground. It was laborious and painstaking work to isolate even the smallest quantities of GFP. Because it was easier, but by no means simple, to isolate aequorin, and it was the protein producing the light, Shimomura focused his research efforts on aequorin and relegated GFP to side projects. (8) In 2008, Shimomura reminisced, "GFP is highly visible and can be easily crystallized, however the content of the GFP in the jellyfish was extremely low, much lower than aequorin. Therefore, to study GFP we had to accumulate GFP little by little for many years while we studied the mystery of aequorin bioluminescence." (9, p. 5600)

Aequorin itself has very interesting and useful properties. In 1967, Ellis Ridgway and Christopher Ashley, both at the University of Oregon at the time, injected the aequorin isolated from 10,000 jellyfish into the muscle fiber of a barnacle. They had read Shimomura's papers and were using aequorin as a calcium sensor because it gives off sparkling blue light in the presence of calcium ions. When they zapped their injected fiber with electric current, it contracted and lit up. (10) Their experiment was the first to prove that calcium was released during muscle contraction and the first to ever use jellyfish proteins to observe the chemical changes occurring inside a living organism. Four years later, Ridgway collaborated with Alan Hodgkin at Cambridge University. This time they injected aequorin into giant neurons of a squid and showed that calcium enters the neurons during neuronal firing. (11)

Ridgway and Ashley's experiments laid the foundations for decades of research in neurobiology and were very important. They had to be, because each injection required the aequorin of thousands of jellyfish. For the technique to be generally useful, a way around the tedious aequorin isolation and jellyfish collection was required.

Despite aequorin being the main focus in his lab, Shimomura still managed to lay the foundation for the GFP revolution. Nearly all of the early GFP knowledge came from his lab. But he was not perfect, and he made one mistake that had far-reaching consequences. When he isolated GFP, there were no other known organisms that had proteins that could fluoresce by themselves. Shimomura therefore assumed that *A. victoria* had some enzymes, found only in the crystal jellyfish, that changed a non fluorescent precursor of GFP into the final fluorescent GFP found in its photoorgans.

Science can be cruel. The difference between working at a $10-an-hour job at a car dealership and winning the Nobel Prize with its associated fame and $1.2 million prize can be smaller than the width of a hair, as was the case for Douglas Prasher. In 2008, Osamu Shimomura, Martin Chalfie, and Roger Tsien won the Nobel Prize in Chemistry for their discovery and development of the green fluorescent protein. In interviews following the announcement, both Chalfie and Tsien acknowledged Douglas Prasher's contribution to their research (12), but few people know Prasher's story and how close he came to getting the hundredth Nobel Prize in Chemistry.

Douglas Prasher was born in 1951 in Akron, Ohio, on the other side of the world from Osamu Shimomura. His father worked at the Goodyear tire factory, but Prasher knew that his ambitions lay elsewhere. Elsewhere turned out to be a PhD from Ohio State University in 1979 and a subsequent postdoctoral assistantship with Milton Cormier at the University of Georgia, who was studying the light given off by the sea pansy and had an interest in the crystal jellyfish bioluminescence. Together with Rick McCann, a technician in the lab, Prasher and Cormier found the gene for aequorin in the crystal jellyfish. They were able to copy it and place it into bacteria so that the bacteria would express the aequorin themselves. Suddenly, if you wanted to produce aequorin, it was no longer necessary to go to the Pacific Ocean to catch a few thousand jellyfish and painstakingly isolate a few thousandths of a gram of aequorin; instead, all that was needed were the genetically modified bacteria that could produce aequorin in large, extractable amounts. This not only saved the lives of millions of jellyfish but also made aequorin much more accessible and easy to study. (13)

After his postdoc, Prasher got a tenure-track job at Woods Hole Oceanographic Institute. There he wanted to use his bacterially produced aequorin to study its photochemistry, but he also had another radical idea. Prasher thought there was a possibility that Shimomura was wrong about GFP needing an additional protein to become fluorescent. This meant that it would be possible to utilize GFP's fluorescence as a tag that would light up when a protein was made and show where it moved in a cell or even in a whole organism. Proteins are very small, and it is impossible to see them in a cell, let alone track their movement in a living organism. By attaching GFP to a protein, one could make a modified protein that fluoresces, which would be much easier to detect: it's like seeing the light of a firefly even if you are too far away to see the firefly itself.

"It was a very risky project and it was very difficult to find funding for the project. I knew that the technique would be incredibly useful if it worked, but it was hard to convince my colleagues of this," Prasher told me. No other fluorescent proteins were known at the time, and it was hard to persuade colleagues and funding agencies that GFP was inherently fluorescent and that it didn't need a special jellyfish enzyme to make its fluorophore. (Scientists working in this area of research call the part of the protein responsible for the fluorescence the "chromophore" or also "fluorophore".)

Douglas Prasher wrote numerous proposals. Finally, one to the American Cancer Society was funded with $200,000. The basis of his proposal was fairly simple. One can think of DNA as a recipe book. At birth every cell has a complete recipe book; it will never get a new one, so the book is kept in a very safe place, in the nucleus of each cell. The recipe book contains the instructions (genes) that describe how to make all the different proteins the body requires. When a new protein needs to be made, a promoter finds the appropriate gene, the instructions are copied until the end of the gene is reached where there is a stop codon that indicates that this is the end of the gene. Once the complete gene has been read, the protein recipe is sent as a messenger RNA to the ribosome, where the protein is expressed. Making copies of the gene in the nucleus and then sending them to the ribosome might seem rather laborious, but it protects the recipe book from the harsh chemical environment of the cell. The process is rigidly controlled. For example, although the genes with the instructions for producing proteins required in the eye are found in every cell in the body, they are copied and translated only in the eye, and only when they are needed. If there is a breakdown in this process or if the recipes are corrupted, sickness and disease follow. Prasher proposed taking the GFP gene and inserting it at the end of the gene of an interesting protein, just before the stop codon. In this way, the protein of interest would be made with GFP tagged onto its end, and the green fluorescence would show when and where the protein was made. It is a bit like going to your favorite recipe and adding the following sentence at its end: "Now ring the bell until I come." You have modified the recipe so that you will always know when your favorite food has been made.

If Prasher's idea worked, scientists would have a noninvasive way of seeing what was happening inside a live cell. They would be able to see proteins moving and reacting in live organisms, and it would completely change the way science was done.

At the time Prasher submitted his proposal, the location of the GFP gene in the jellyfish genome was unknown. Shimomura had isolated GFP from jellyfish, but he wasn't interested in its gene. This meant that Prasher had to catch some jellyfish of his own so that he could extract their RNA and search for the GFP gene and clone it. Today it would be fairly easy to find the gene for a protein like GFP in a jellyfish. It would take a week or two and require just a couple of jellyfish. In the early 1980s, when Prasher was doing most of his research, however, the molecular biological techniques commonly used today were unknown, and it was an extremely difficult process. Finding and isolating the gene took longer and required more jellyfish, especially because on his first attempt he only got half the GFP gene and he had to return to Friday Harbor to get more jellyfish. (3, 4)

One of the few other researchers to show any interest in Prasher's GFP work was Marty Chalfie, a biologist at Columbia University. In 1989, while attending a seminar on bioluminescence, Chalfie heard about GFP for the first time and had the same idea as Prasher; he wanted to use GFP to show when proteins are made, particularly in the transparent roundworm, *Caenorhabditis elegans*, which he was using as his research model. Describing his introduction to GFP, Chalfie said, "Every one of my seminars, I would bore people by listing reasons why you would want to work with this worm. I would go through the list and at one point or another I would say that the animals are transparent. As soon as I heard about this wonderful protein, I was primed to fantasize about using GFP to see when proteins were made in our worms."

After the seminar, Chalfie learned about Prasher's work and called him to see how far along in the project he was and to suggest a collaboration. Because *C. elegans* are transparent, and Chalfie was undoubtedly one of the world's *C. elegans* experts, Prasher was very interested in working together. Unfortunately, Chalfie's call to Prasher was a year too early; Prasher had just discovered that he had only cloned and sequenced a fragment of the GFP gene. Prasher promised to send Chalfie the gene once he had properly isolated it in its entirety.

Over a 3-year period, Prasher caught roughly 70,000 jellyfish, which gave him enough messenger RNA for two rounds of gene searching. Prasher's perseverance eventually paid off, and he finally found and sequenced the complete GFP gene. He inserted the gene into *Escherichia coli*, which then produced the protein. Sadly, the GFP did not fluoresce. Douglas Prasher was devastated but not greatly surprised. It seemed that just as Shimomura had predicted, the *E. coli* needed an additional jellyfish enzyme to help them modify GFP into its fluorescent form. Because his money for the project had run out, Prasher applied for a renewal of his grant. Unfortunately, the American Cancer Society did not approve his proposal for additional funding to complete the GFP tagging project. Dejected, Prasher gave up on the project. Like Shimomura, he decided to ignore GFP and focus on aequorin. As promised, Prasher also tried to get in touch with Marty Chalfie so that he could send him the complete gene. Unfortunately, Chalfie was on sabbatical at the University of Utah, and they failed to make contact. A few years later, Prasher told me why he didn't persevere with his GFP project: "I got no support for my ideas from my colleagues, and by the time I got the *E. coli* to express GFP, I was already doubting myself and was not expecting to observe any fluorescence."

Indeed, besides the one phone call from Chalfie, Prasher did not find much interest or support for his research. His colleagues were skeptical that the project would work, and some were even dismissive of the research. This was in part due to the fact that Prasher was one of the first molecular biologists working at Woods Hole Oceanographic Institute and had no natural colleagues. He was also neither very outgoing nor a natural salesman.

Scientists must have more than their fair share of self-confidence if they want to change the way in which science is done. The scientific method requires that researchers publish their ideas and have them withstand the rigors of peer review. This is a scary process, especially when the ideas are so radical that there is little support for them in the community. Believing in one's work and having the courage to fail are critical requirements for paradigm shifts in science. Prasher's idea had the same import as those put forth by Galileo, Salk, and Watson and Crick, and his experimental technique was flawless; he just lacked the drive, confidence, and doggedness to plow through the obstacles in his way.

Shimomura and Prasher were colleagues for all of Prasher's tenure at Woods Hole, yet the only time they spoke to each other was at a bioluminescence conference held thousands of miles away, at Friday Harbor. With a little more support, Prasher might have persevered with GFP and made the breakthrough that he was so close to achieving.

Before closing shop on GFP, Douglas Prasher published a paper in *Gene* in 1992, describing how he isolated the GFP gene and giving its sequence. He wrote, "These results will enable us to construct an expression vector for the preparation of non-fluorescent apo-GFP." (*14*, p. 233) Thus, he was reiterating his belief that the genetically modified bacteria could make GFP, but that it didn't have the machinery to make it glow—machinery that was found only in jellyfish (an apoenzyme is an enzyme that is missing its prosthetic group).

Marty Chalfie didn't read Prasher's paper when it first appeared. It was the arrival of Ghia Euskirchen, a rotation student in his lab who had studied fluorescence while working on her master's thesis in chemical engineering, that prompted him to find out if any other fluorescent proteins besides GFP had been discovered. In the process he found the *Gene* paper and contacted Prasher, requesting the GFP gene. Prasher felt guilty that he hadn't persevered and gotten hold of Chalfie earlier, so now he sent him the gene. Many years later, Prasher would tell the *Huntsville Times*, "I could have hung onto the gene, but when you're in that environment and losing public funds, you've got an obligation to share." (*15*)

Chalfie received the GFP DNA in September 1992 and entrusted Ghia Euskirchen with the GFP project. When Prasher had isolated and copied the GFP gene, he had copied 25 additional base pairs before the gene and 227 after it, which was easier than isolating just the gene itself. It was common practice at the time, and he didn't think that it would make any difference. In trying to add the GFP gene to *E. coli* so that they would express glowing GFP, Euskirchen made one big change to Prasher's procedures. She used primers that copied only the coding sequence so that she could use the polymerase chain reaction (PCR), a very new technique at the time, to copy the gene before inserting it into *E. coli*. PCR enables researchers to accurately make millions of copies of a specific DNA sequence in just a few hours. According to its inventor, Kary Mullis, "PCR can generate 100 billion similar molecules in an afternoon. The reaction is easy to execute. It requires no more than a test tube, a few simple reagents, and a source of heat." (*16*, p. 56) It made Euskirchen's job much easier and had a very fortuitous consequence.

Using the fluorescence microscopes in the chemical engineering department at Columbia University where she had done her master's work, Ghia Euskirchen was the first person in the world to see GFP expressed in *E. coli* fluoresce (figure 1.3). In Prasher's lab, that little bit of extra DNA before and after the GFP gene, much shorter than the width of a hair, was responsible for the GFP folding incorrectly in the bacteria, and it prevented GFP from emitting its signature green glow! Euskirchen had overcome the final hurdle by amplifying the DNA with PCR and not using restriction enzymes. GFP expressed in *E. coli* was fluorescent; it didn't need an additional jellyfish protein. (*17*)

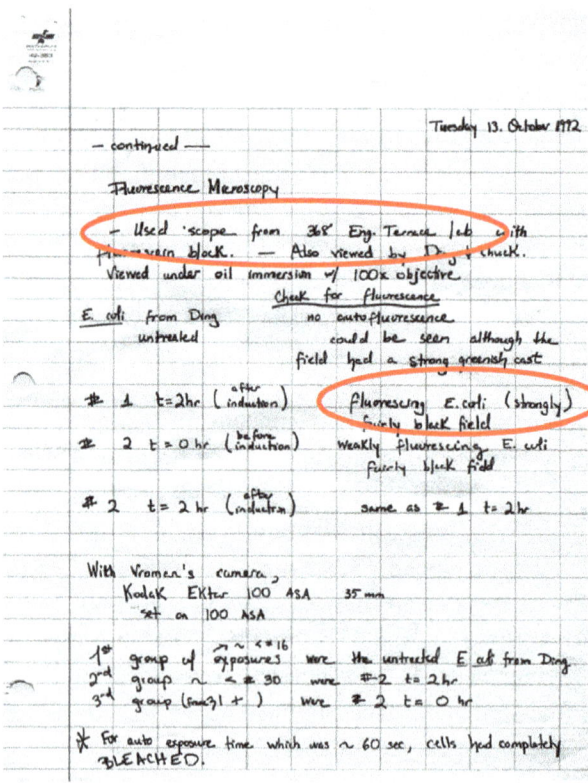

Figure 1.3 Ghia Euskirchen's lab notebook for October 13, 1992. (*12*)

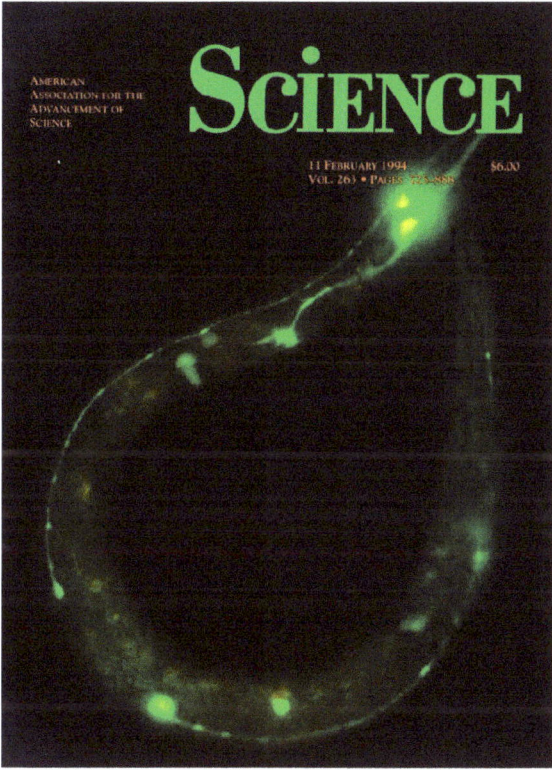

Figure 1.4 The cover of the February 11, 1994, issue of *Science* shows a genetically modified *C. elegans* with GFP expressed in the touch receptor neurons. The photograph came from the breakthrough paper by Martin Chalfie that described the first use of green fluorescent protein as a protein marker. (*19*)

Chalfie studies *C. elegans*, and he was particularly interested in how the transparent roundworm responds to touch. In 1993, his lab replaced the gene for a touch-sensitive protein with that of GFP. The paper describing the work was accepted in *Science*, and the image of the genetically modified *C. elegans* associated with the manuscript was reproduced on the front cover of the magazine (figure 1.4).

Chalfie often jokes that he had to do a lot of behind-the-scenes negotiations to get the paper and the magazine cover to look the way he wanted. First of all, green photographs are difficult to reproduce and do not make good covers, so Chalfie had to work hard to convince the *Science* art director that a color change would be inappropriate for the cover. Second, Chalfie wanted to cite the unpublished work of Tulle Hazelrigg, who was the first person to express GFP in fruit flies. However, on November 11, 1993, Hazelrigg, a professor of biology at Columbia University, sent the following formal reply to Chalfie's request: "It is perfectly fine with me if you cite S. Wang's and my unpublished results in your *Science* paper on GFP, provided you meet the following conditions: 1. You make coffee each Saturday morning for the next two months, ready by 8:30 a.m. 2. You prepare a special French dinner at a time of your choosing. 3. You empty the garbage nightly for the next month." Chalfie agreed to her terms and was allowed to cite the work done by Tulle Hazelrigg, who happens to be his wife, in the *Science* paper. There is some debate regarding whether he actually lived up to his agreement. (*18*)

All joking aside, Hazelrigg's work is very important, and it deserved a citation in Chalfie's *Science* paper (*19*) and in any discussion of fluorescent proteins. Hazelrigg was the first person to tag a protein with GFP. Her paper describing the work was published in *Nature* in 1994, and it was the second recombinant GFP publication ever. Scientists now had a new tool that would open windows to worlds barely imagined before. However, like a car standing on the top of a hill, this revolution needed a push to get it going.

Roger Tsien was the man to give GFP technology the shove it needed. He had always been interested in imaging technology and was well known in the field for his work on calcium sensors. Upon reading Prasher's *Gene* paper, he realized the potential of GFP in protein imaging. Tsien requested a copy of the GFP gene. He wanted to express GFP in his own lab to be sure that Prasher was correct that GFP wouldn't fluoresce outside the jellyfish. If it was indeed nonfluorescent outside the crystal jellyfish, as Tsien believed, he wanted to find the jellyfish protein that was required to make GFP fluorescent.

Roger Tsien, who won the Westinghouse Science Talent Search as a 16-year-old, was a chemist, so when Prasher sent him the GFP gene by mail, he had no one in his lab capable of working with it. He was waiting for Roger Heim, a new postdoctoral associate, to arrive in his lab. Heim had experience with molecular biological techniques and would be well prepared to work with GFP. However, before Heim got to the lab, Tsien heard that Chalfie had beaten him to the punch and had expressed GFP that fluoresced. Rather than repeat Chalfie's work or even overlap with it, Tsien and Heim struck out in a new direction. "I heard that bacteria worked, so we set out do yeast, specifically not to be treading on Marty's toes," Tsien said in an interview given at the Nobel Prize awards ceremony. (*20*) Very soon after, Heim, and Tsien expressed GFP in yeast and modified it so that it fluoresced blue, yellow, and cyan instead of green (figure 1.5). They also created the basis of a much brighter version of GFP, called enhanced *GFP* (EGFP). (*21*)

Figure 1.5 An agar plate of bacterial colonies expressing differently colored fluorescent proteins. These fluorescent proteins developed by Roger Tsien's group are called the mFruits and have names like mHoneydew, mTomato, mCherry, mRaspberry, and mPlum. (Courtesy of Paul Steinbach and Roger Y. Tsien, University of California, San Diego.)

EGFP is still the most commonly used fluorescent protein, even though today there are many commercially available fluorescent proteins. So many that for most experiments it is quite an art to pick the right one. As Roger Tsien says, "People are developing more and more colors. It's quite a confusing zoo for those who are not in the know. It is almost like buying a car now, you have to keep track of an awful lot of models."

The fluorescing *C. elegans* on the cover of *Science* magazine attracted the attention of a number of scientists, including Robert Hoffman, president of AntiCancer Inc., in our discussions he said, "When I saw Chalfie's worm on the cover of *Science*, I got excited. At that time we were very interested in not only cancer but also gene therapy and hair growth. My first thought was to see if we could use GFP as a model gene for delivery into the hair follicle." The AntiCancer group managed to do that, but more significantly, it would also use GFP to light up cancer cells, thereby becoming the first researchers to illuminate diseases with fluorescent proteins. A little later AntiCancer Inc. would make history again by showing how GFP could be used to monitor the progression of a disease in a live animal in real time.

However, nothing in science occurs in isolation. At the same time as the GFP technology was being developed, a parallel process was occurring in the world of firefly luciferase.

Male fireflies use their light organs to attract females. Males from every species of firefly emit a unique sequence of light flashes, allowing them to court only females from their own species. The light is produced in the light organs located in the posterior of the firefly, when a small luciferin molecule binds to the protein luciferase. Although luciferase can be used in a similar manner to GFP, luciferase does not fluoresce; instead, it gives off light. All the energy required for the light flash comes from ATP, an energy-rich molecule. Luciferase is an enzyme whose job it is to hold the luciferin, ATP, and oxygen in close proximity to each other so they can react. Once they have reacted and flashed their light, the luciferase is ready to take a new oxygen molecule, ATP, and luciferin and make them react again. If the poor firefly is crushed with a pestle and mortar, all four components will mix, and the resulting firefly mush will glow until one of the four components runs out. All of this occurs without any increase in heat; it is cold light.

Transgenic organisms that produce luciferase can be created just like the GFP-producing *E. coli* and *C. elegans* described previously. However, cells or organelles that express luciferase are dark until they are injected with luciferin. There is no need to inject ATP because it is a ubiquitous energy source present in all living cells. The advantage of using luciferase is that the light comes from the modified cells; the disadvantage is that luciferin has to be injected into the cells before the luminescence is produced. GFP labeling, on the other hand, doesn't require injection of any substrates, but GFP needs to be excited before it fluoresces, which means that some light has to penetrate to the genetically modified cells.

Bruce Branchini, my colleague at Connecticut College, has been working with firefly luciferase for many years, and in 1994, he invited Doug Prasher to give a talk at the College. Branchini wanted to hear all about the jellyfish protein that was threatening to upstage his firefly lights. As a junior faculty member, I was in charge of the slide projector. I had never heard of GFP, which was a relatively unknown molecule at the time. In fact, only four papers on GFP had been published. Nearly 20 years later, I can still remember standing at the back of the lecture room looking over no more than 15 restless undergraduates. It was not the most impressive crowd, and the tall, balding speaker struggled to hold the students' attention. He was not a natural speaker, and the importance of the material was lost on the listeners. Consequently, when Prasher stressed that GFP had the potential to be big—very big—in the conclusion of his talk, I wasn't quite convinced. However, GFP did sound interesting, and so I started my academic career working on a protein that was about to become an indispensable tool in the arsenal of biomedical labs all over the world. At the time, fewer than 20 people were studying fluorescent proteins, but now more than 3 million experiments each year utilize these illuminating proteins to show us when and where proteins are made in living organisms, thereby increasing our knowledge of the mechanics of disease and providing new avenues in the search for their cure. In the prologue to a 2004 book describing methods and techniques for using fluorescent proteins, Martin Chalfie wrote that in the prior year more than 60 percent of the papers published in the *Journal of Cell Biology* and more than 50 percent of those published in *Cell* mentioned fluorescent proteins. The field had lived up to Prasher's promise.

Because there was a need for a central site that could introduce newcomers to the utility and beauty of fluorescent proteins, I started collecting information on interesting uses of GFP and began a GFP website. Our website became increasingly popular with high school teachers and apparently, also with the Royal Swedish Academy of Sciences. On November 16, 2007, I walked into my office after having taught an introductory chemistry class at 8:00 a.m. The light on my phone was flashing. Intrigued, because I seldom get messages that early in the day, I immediately called my voicemail. The message was from Professor Gunnar von Heijne calling on behalf of the Royal Swedish Academy of Sciences and the Nobel Committee for Chemistry, asking me to please return his call because he wanted to talk to me about GFP. The committee invited me to Stockholm and requested background information on everyone's individual contributions to the development of the GFP technology.

Although I am not allowed to reveal the details of my trip, I can say that I was extremely impressed by the committee's knowledge of the ins and outs of the GFP story. Our conversation was just a small part of its larger investigation. The Nobel Prize cannot be awarded to more than three people. The committee members must have had some difficult and intense debates in deciding between the four scientists who gave birth to the GFP revolution: Shimomura, Prasher, Chalfie, and Tsien.

On October 8, 2008, the Nobel Foundation was scheduled to announce the chemistry prize, so at 5:30 a.m. my wife and I were in front of the computer watching the live broadcast of the award and heard the exhilarating news: the hundredth Nobel Prize in Chemistry was awarded to Osamu Shimomura, Martin Chalfie, and Roger Tsien, "for the discovery and development of the green fluorescent protein, GFP."

Prasher, whose seminar was the impetus of my interest in GFP, was not among the prizewinners. His near miss motivated some journalists to solicit comments from him. It was then that they discovered Doug Prasher's story is a tragic one, for not only did he miss the Nobel Prize in Chemistry by 25 nucleotide bases, less than the skin of his teeth, but life had been unkind to him.

As a biochemist, Prasher felt unappreciated at Woods Hole, with its strong focus on marine biology, and shortly after giving up his quest for a fluorescent protein tag, he left the institute for a position researching plant pests at the US Department of Agriculture. The job turned out to be more stressful than he had expected, and at 51 Prasher had a minor heart attack. This led him to AZ Technology, a NASA subcontractor located in Huntsville, Alabama. While there, Prasher enjoyed developing handheld sensors designed to detect harmful bacteria in spacecraft. He liked living in Huntsville, and so he was devastated when NASA cut the funding for the handheld sensor project, and his job was made redundant yet again. Because he didn't want to move his family from Huntsville, Prasher spent a year looking for employment in the local science sector. There weren't many opportunities, and he failed to land one of the positions he applied for at the Hudson Alpha Institute for Biotechnology and Calhoun Community College. After a year of looking for a position appropriate for a PhD biochemist, Prasher's finances and his need to productively occupy himself forced him to search for a job outside the sciences. When he drove by Bill Penney's Toyota, which is ironically located on University Drive, and saw that it was looking for a courtesy van driver, he applied and became an overqualified shuttle bus driver. That is the job he had when he heard that the 2008 Nobel Prize in Chemistry was awarded to Shimomura, Chalfie, and Tsien. (22)

In interviews following the announcement, both Chalfie and Tsien were magnanimous and repeatedly acknowledged Douglas Prasher's contribution to their research. Marty Chalfie has publicly stated that he wished that there was not a three-person limit on the number of scientists who can win the Nobel Prize in each of the categories. "Cloning GFP was essential to this entire project," said Chalfie. "Without it, neither my work nor Roger's work would have been possible." (23, p. 16) Tsien and Chalfie invited Prasher and his wife, Virginia, to the Nobel ceremonies and paid for their flights. Roger Tsien also offered Prasher a job in his laboratory in San Diego. At first, Prasher did not want to leave Alabama, so he did not accept Tsien's offer, but in 2012 the lure of doing productive research won Prasher over. He finally made the move, and he now works in Roger Tsien's lab.

A few weeks after the prize was announced, an invitation for my wife and me to attend the Nobel Prize award ceremony and banquet arrived in my campus mailbox. It came with a major incentive, not that we needed one—a week´s accommodation at the elegant Grand Hotel, where together with friends and relatives of each Nobel laureate we would experience Nobel week in Stockholm.

Nobel week was magical. The December sun sets at 2:45 p.m. in Stockholm, and the paths to the Nordic Museum and the Royal Swedish Academy of Sciences, sites of some of the receptions, were lit up with flickering torches. Inside the beautifully decorated halls were equally decorated laureates, their friends, and members of the academy. Here we sipped champagne and enjoyed reindeer pâté. Before coming to Stockholm, I knew three of the Nobel laureates. I had met Luc Montagnier, the discoverer of the AIDS virus, since he was awarded an honorary doctorate at Connecticut College; and I knew Chalfie and Shimomura through our common interest in GFP. However, the laureates were in demand at the receptions, and we spent our time talking with members of the selection committee and GFP friends, including Euskirchen and Prasher, who were there as Tsien and Chalfie's guests. Twice we had breakfast with Doug, he was enjoying the festivities and the sights of Stockholm. There was no outward indication of the hurt and disappointment he must have been feeling.

To me, the unquestionable climax of the week was the award ceremony and banquet hosted by King Carl XVI Gustaf and Queen Silvia. It was an intellectual hybrid of the Olympic opening ceremonies and the Academy Awards, broadcast live on Swedish and Japanese television. The menu and the dresses worn by the queen and princesses were well-kept secrets until the evening of December 10, the anniversary of Alfred Nobel's death. The next day they were all over the Internet and on the front page of the Swedish dailies. Chalfie was seated beside Crown Princess Victoria, while economics laureate Paul Krugman enjoyed the company of her younger sister, Princess Madeleine. Prince Carl Philip had Wendy Tsien, wife of chemist Roger Tsien, as his guest of honor.

On the morning of the feast, more than 7,000 porcelain pieces, 5,000 glasses, and 10,000 pieces of silverware were meticulously laid out on the 1,500 feet of linen that adorned the banquet's 65 tables. The spectacle surrounding the meal must be part of the reason the Nobel banquet is an annual TV viewing tradition in Sweden. Accompanied by a fanfare, waiters processed down a central staircase balancing the food above their heads. The king was served first, immediately followed by the queen. The desserts arrived in the dark, each waiter carrying three servings of pears Hélène with flares shooting out 3 feet of sparks. To me, the highlight of the evening was Tsien's banquet speech in which he thanked the corals and jellyfish for their shine. (24)

The 2008 Nobel Prize in Chemistry rewarded both basic and applied research. While Shimomura's primary interest in GFP was its role in *Aequorea* bioluminescence, Tsien was interested in its practical applications. He developed brighter, faster-maturing, more photostable, and multicolored fluorescent proteins, then incorporated them into in vivo sensors. Fluorescent proteins demonstrated that basic research can open the doors to very useful and often unexpected discoveries. There is clearly room for both basic and applied research. Roger Tsien eloquently stated the case for applied research in his Nobel speech: "Some people have at times criticized us for mainly working on techniques. I would like to draw their attention to an old Chinese proverb that says that if you give a man a fish you feed him for one day, if you teach him how to fish you feed him for a lifetime. That's why we enjoy devising fishing tackle and nets to scoop from the ocean of knowledge." (25, p. 2831) During the Nobel awards banquet in 2008, Roger Tsien, speaking for the three chemistry award winners, also said, "We hope this prize reinforces recognition of the importance of basic science as the foundation for practical benefits to our health and economies." (26)

Green fluorescent protein has been floating in the ocean for more than 160 million years, but it took a curious scientist, fascinated by pinpricks of green light, to begin unlocking its potential. Now, GFP is the microscope of the twenty-first century. In technicolor, it lets us see things we have never been able to see before, thereby completely changing the way we approach science and medicine. And its reign has only just begun. "We are entering an era where the blots, the gels, the radioactivity, and the twentieth-century molecular biology will be replaced by real-time imaging in live animals. I see the day when there will be transgenic people, not for movies, but with fluorescent proteins to report the onset of disease," said Robert Hoffman of AntiCancer Inc.

> While environmentally friendly or so-called "green" chemistry has become all the rage in the chemical community, no human chemist can yet match what a single jellyfish gene directs: 238 ordered condensations + 1 cyclization + 1 oxidation, all done in a few minutes in aerated water with no protecting groups, only one slightly toxic by-product, and essentially 100 percent yield of an extremely useful product that literally glows green. Corals produce yellow and red FPs with the same chemistry plus one extra oxidation. Yet coral reefs are also under worldwide jeopardy, due to acidification and warming of the oceans. So my final thanks are to both the jellyfish and corals: long may they have intact habitats in which to shine!
>
> Roger Tsien, speech, 2008 Nobel Prize award banquet

REFERENCES

1. Shimomura, O. (2005). The discovery of aequorin and green fluorescent protein. *Journal of Microscopy 217*, 3–15.
2. Osamu Shimomura biography. http://www.nobelprize.org/nobel_prizes/chemistry/laureates/2008/shimomura-bio.html.
3. Zimmer, M. (2005). *Glowing genes: A revolution in biotechnology*. Amherst, NY: Prometheus Books.
4. Pieribone, V. A., and Gruber, D. F. (2005). *Aglow in the dark: The revolutionary science of biofluorescence*. Cambridge, MA: Belknap Press of Harvard University Press.
5. Zimmer, M. (2014). Introduction to fluorescent proteins. In R.N. Day and M.W. Davidson, Eds., *The fluorescent protein revolution*. (pp. 3–24) New York: CRC Press.
6. Shimomura, O. (2003). Personal interview.
7. Shimomura, O., Johnson, F. H., and Saiga, Y. (1962). Extraction, purification and properties of aequorin, a bioluminescent protein from the luminous hydromedusan, Aequorea. *Journal of Cellular Comparative Physiology 59*, 223–239.
8. Shimomura, O. (1998). The discovery of green fluorescent protein. In M. Chalfie and S. Kain, Eds., *Green fluorescent protein* (pp. 3–15). New York: Wiley-Liss.
9. Shimomura, O. (2009). Discovery of green fluorescent protein (GFP) (Nobel Lecture). *Angewandte Chemie—International Edition 48*, 5590–5602.
10. Ridgway, E. B., and Ashley, C. C. (1967). Calcium transients in single muscle fibers. *Biochemical and Biophysical Research Communications 29*, 229–234.
11. Baker, P. F., Hodgkin, A. L., and Ridgway, E. B. (1971). Depolarization and calcium entry in squid giant axons. *Journal of Physiology 218*, 709–755.
12. Martin Chalfie interview. http://www.nobelprize.org/nobel_prizes/chemistry/laureates/2008/chalfie-interview.html.
13. Cormier, M. (2008). Cloning the GFP gene. *American Society for Photobiology News 39*, 4–5.
14. Prasher, D. C., Eckenrode, V. K., Ward, W. W., Pendergast, F. G., and Cormier, M. J. (1992). Primary structure of the Aequorea victoria green fluorescent protein. *Gene 111*, 229–233.
15. Doyle, S. (2008). Local biochemist had hand in Nobel. *Huntsville Times*, 10 October.
16. Mullis, K. B. (1990). The unusual origin of the polymerase chain-reaction. *Scientific American 262*, 56–61.
17. Euskirchen, G. (2008). Interview with author.
18. Chalfie, M. (2009). GFP: Lighting up life (Nobel Lecture). *Angewandte Chemie—International Edition 48*, 5603–5611.
19. Chalfie, M., Tu, Y., Euskirchen, G., Ward, W. W., and Prasher, D. C. (1994). Green fluorescent protein as a marker for gene expression. *Science 263*, 802–805.
20. Roger Tsien interview. http://www.nobelprize.org/nobel_prizes/chemistry/laureates/2008/tsien-interview.html.
21. Heim, R., Cubitt, A., and Tsien, R. Y. (1995). Improved green fluorescene, *Nature 373*, 663–664.
22. Bhattacharjee, Y. (2011). How bad luck and bad networking cost Douglas Prasher a Nobel Prize. *Discover Magazine,* April 2011.

23. Talan, J. (2008). Why a little green signaling protein prompted this year's Nobel Prize in Chemistry—The Nobelists share their story. *Neurology Today 8*, 15–16.
24. Zimmer, M. (2009). In the glow of Alfred Nobel. *Connecticut College Magazine*, Spring, 20–23.
25. Zimmer, M. (2009) GFP: from jellyfish to the Nobel prize and beyond. *Chemical Society Reviews 38*, 2823–2832.
26. Roger Y. Tsien banquet speech. (2008). nobelprize.org.

CHAPTER 2

Heart Disease

Heart Disease in the United States, 2010	CDC (1)
Noninstitutionalized adults with diagnosed heart disease	27.1 million
Hospital discharges with heart disease as first-listed diagnosis	4.0 million
Average length of stay	4.6 days
Number of deaths	599,413
Cause of death rank	1

> Thus nature, ever perfect and divine, doing nothing in vain, has neither given a heart where it was not required, nor produced it before its office had become necessary; but by the same stages in the development of every animal, passing through the forms of all, as I may say (ovum, worm, foetus), it acquires perfection in each. These points will be found elsewhere confirmed by numerous observations on the formation of the foetus. Finally, it is not without good grounds that Hippocrates in his book, "De Corde," entitles it a muscle; its action is the same; so is its functions, viz., to contract and move something else—in this case the charge of the blood.
>
> William Harvey, *On the Motion of the Heart and Blood in Animals* (1889)

How is that reassuring rhythm of a parent's or lover's heartbeat regulated? Is there a clock or some other internal device that is responsible for the timing of our heartbeat? Yes, there is a protein that controls the fundamental short-term rhythms, called *ultradian rhythms*, in our bodies. The protein regulates the contractions of smooth muscles by regularly raising and lowering calcium levels in the muscle cells. The illumination of the rhythm protein is one of my favorite uses of fluorescent proteins because it is a beautiful example of starting with questions about what is occurring in the human body, using a model organism to answer the questions, and then extrapolating back to humans. And it involves food, sex, and saliva, which can never be bad.

Andres Villu Maricq, a biology professor and physician at the Center for Cell and Genome Science at the University of Utah, has been interested in what makes our hearts tick for quite a while. He had a hunch that he knew which protein was the internal rhythm keeper in humans. However, it is very difficult to study the effects and workings of a protein in humans. Not only are there numerous ethical and regulatory problems, but people are complicated, with many thousands of proteins. A much simpler organism was needed to study the workings of the rhythm protein, the roundworm *Caenorhabditis elegans* seemed like a great choice. In 1998 its genome was determined, making it the first multicellular organism to be completely sequenced. It is a great model system to study human diseases because, although the worm has only 959 cells about 65 percent of human disease genes have a counterpart in *C. elegans*. This led Bruce Alberts, president of the National Academy of Sciences, to remark: "We have come to realize humans are more like worms than we ever imagined." (2, p. A1)

Maricq went to the worm genome to find out whether it had a protein equivalent to the human rhythm protein. He and his colleagues scanned through the roughly 20,000 genes in *C. elegans* and found a gene that was very similar to the genes for three human timekeeper proteins. Its function in the worm was unknown at the time, so the Utah scientists created some transgenic worms with the green fluorescent protein (GFP) gene tagged onto the gene of the putative rhythm gene. The resultant worms emitted green fluorescence from their throats, intestines, and reproductive organs. The protein is crucial to the worm's existence; eliminating it resulted in death. But by selectively restoring the gene to certain areas of *C. elegans*, Maricq was able to tease out the function of the rhythm gene at each of the three separate locations.

The throat of a normal worm undergoes a rhythmic, swallowing-like motion every quarter of a second. Without the rhythm protein in their throats, the worms could not swallow their food or saliva, and they died before reaching the first of their four larval stages. Humans produce more than a liter of saliva a day; without knowing it, we are constantly swallowing, and if we didn't we would drown in our own spit. It would be interesting to know whether an analogous rhythm protein controls this swallowing in *Homo sapiens*.

When the researchers in Maricq's laboratory switched off the light in the worms' intestines, they created constipated worms. Instead of having regular bowel movements every 45 to 50 seconds, the worms that were not expressing the timer protein around the intestine had average times of more than 3 minutes between defecations and never grew to full size (figure 2.1). "The worm can swallow and live and grow up to be an adult with really bad constipation," Maricq says. "Mutants sometimes have to wait six to 10 minutes." (3,4)

Most of the volume of the worm is taken up by its reproductive system. C. elegans are hermaphrodites, and only about 0.05 percent of normal laboratory populations are males. The worms frequently have sex, but it doesn't seem to be very exciting. During their self-fertilization, the gonadal sheath contracts every 7 seconds, squeezing out eggs so that sperm can fertilize them. Sperm is the limiting reagent; it runs out before the eggs do. In a typical C. elegans population, it is the function of the 0.05 percent of males to fertilize the older hermaphrodites that have run out of sperm. When the timer gene in the reproductive area is switched off, the sex life of the worm becomes chaotic. The worm loses its rhythm, and the eggs reaching the sperm are too young or too old to be fertilized, and the worms have virtually no offspring. Interestingly, the worms evolve and compensate by increasing the numbers of males in their populations. (3)

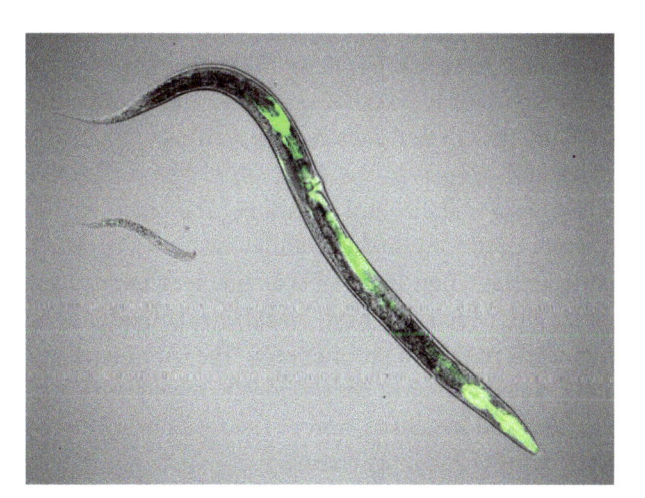

Figure 2.1 A genetically modified *C. elegans* with its rhythm proteins illuminated by GFP. The protein is expressed in the roundworm's throat, intestine, and reproductive organs. When the protein around the intestines is knocked out, the worm becomes constipated and grows to only a fraction of its normal size (see small worm on left). Worms without the rhythm protein in their throat don't swallow regularly and drown in their own spit. (3) (Courtesy of Ken Norman, University of Utah.)

All the rhythms examined in the worm are non-neurogenic, which means that they do not appear to be driven by nervous system input. This makes them good models for the intrinsic rhythms found in tissues such as the mammalian heart. In fact, 2 years after Maricq and colleagues published "the scoop on when worms poop, ovulate and swallow" (4), Xose Bustelo and coworkers at the University of Salamanca in Spain showed that the very same rhythm gene investigated by Maricq plays a crucial role in mice, and that mice that lack the gene suffer from a variety of cardiac diseases, some of which strongly resemble hypertension in humans. (5)

Next time you feel the heartbeat of a loved one, perhaps you will remember that deep inside that person there is a rhythm protein directing the symphony of life, and that it is responsible for the comforting thumps that are giving you pleasure.

Thanks to the guidance of the rhythm protein, the human heart beats 100,000 times a day, pumping 2,000 gallons of blood a distance of 60,000 miles around your body. Of course, we don't have thousands of gallons of blood or tens of thousands of miles of blood vessels in our bodies; the heart pumps such huge quantities because the blood is continuously circulating through the body. When the body is at rest, it takes only 6 seconds for the blood to go from the heart to the lungs and back, only 8 seconds for it to go the brain and back, and just 16 seconds for it to reach the toes and travel all the way back to the heart. (6) The heart pumps oxygenated blood through the aorta, the largest artery, at about 1 mile an hour. By the time blood reaches the much narrower capillaries, it has to slow down, and it is moving at around 40 inches per hour. (6) In this way, it is the job of the heart to supply blood via its arterial system to almost all of the body's 75 trillion cells. Only the corneal cells receive no blood. (6) The blood carries food molecules and oxygen to the cells, where the cells use them as an energy source to power their vital functions. The blood in the veins then carries the waste produced during cellular respiration, which is mainly carbon dioxide, to the lungs. If the cells do not get their oxygen and nutrients, they seize up, which can have disastrous consequences. The brain and the heart itself require a lot of energy. The brain uses up 20 percent of all the oxygen circulating in the body. When we exercise, our muscles need more oxygen and nutrients, and so the heart beats more frequently, pumping more blood through its chambers.

The heart is a truly amazing organ. To live, and maybe love, we need it to be functioning every day and every night of our lives. If something goes wrong with our hearts, we are in trouble. Unfortunately, the stresses on the heart are enormous, and not all hearts are infallible. A range of problems such as infections, diseases of the blood vessels, arrhythmia, and congenital defects can affect the heart. Heart diseases are the leading cause of death in many countries, including the United States, Canada, England, and Japan, where they account for about a quarter of all deaths.

Thanks to experiments with tiny fluorescent roundworms, we know that there is a rhythm protein that regulates the contractions of smooth muscles and that the protein functions by regularly raising and lowering calcium levels in the muscle cells. In the mammalian heart, efficient beating depends on the coordinated release and reuptake of calcium ions from intracellular organelles in millions of cells, at rates between 0.5 and 15 beats per second throughout life, and even subtle dysfunctions of this process can result in cardiac arrhythmias and sudden death.

Michael Kotlikoff, the dean of Cornell University's College of Veterinary Medicine, and his colleagues have used in vivo calcium-sensing fluorescent probes, such as those described in the chapter 9 to monitor calcium levels in the heart muscle contractions of living mice. Their work, published in 2005, was one of the first in vivo calcium imaging papers, and it has expanded our understanding of how the heart develops in living mouse embryos and could improve our knowledge of irregular heartbeats.

In mammals, the heart is the first organ to function, and it starts beating prior to the embryo's full development. Because the mouse heart beats 6 to 10 times per second, imaging the changing calcium levels was quite a technical feat that required a high-speed camera that was cooled to minus 90 degrees Celsius. Figure 2.2 shows a sequence of images of a beating 10-day-old embryonic mouse heart. At this point in development, the heart has only two main parts, the upper and the lower portions. Contractions are initiated by calcium ions that are released in the upper part of the heart, known as the atria. The calcium concentration gradient spreads as a wave across the atrial surface before being dramatically slowed down in the region of the atrioventricular canal that separates the upper and lower regions. Upon leaving the atrioventricular canal, the waves of calcium ions then rapidly speed up, spreading along the ventricular (lower heart) surface (lower images of figure 2.2). The rate of conduction through the specialized cells in the canal was 10-fold slower than through the atrial and ventricular chambers, providing a sufficient delay for effective mechanical pumping. After 13.5 days of development, the two portions of the heart separate into four, and a functional atrioventricular node takes over the function of the atrioventricular canal. By that time, the technique revealed, the specialized cells in the canal have died so that functions are not duplicated. "These cells have to die, because if they didn't the heart would not function properly," said Kotlikoff. (7)

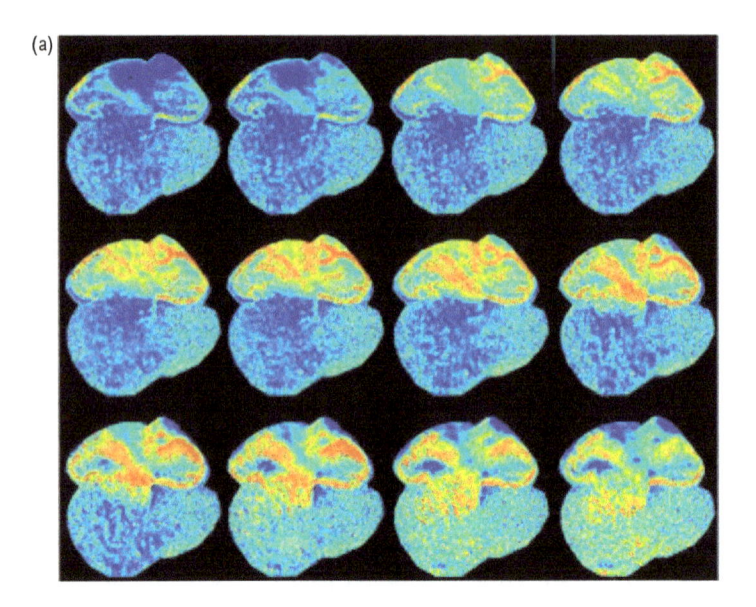

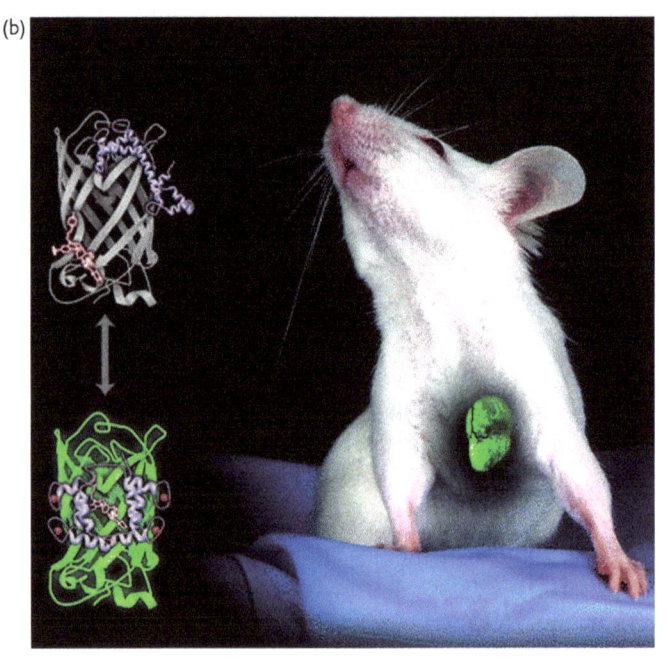

Figure 2.2 a. This series of images reveals increases in cell calcium (yellow through to red) from a mouse embryo's upper heart (atrium through a canal to the lower heart (ventricle) on day 10 of development. Cell calcium rises when muscles contract. (8) b. Transgenic mouse created to illuminate calcium flux in individual heartbeats. GCaMP2 is nonfluorescent in absence of calcium ions (top left) and fluoresces when bound to 4 calcium ions (bottom left) (Photograph and drawings courtesy of Alexis Wenski–Roberts and Michael A. Simmons, MFA.)

Although the paper that was published in the *Proceedings of the National Academy of Sciences of the United States* provided an interesting insight into the beating of developing mammalian hearts, its main importance was that it demonstrated how new and improved calcium sensors can be used. In fact, the authors conclude their article with the following statement: "The rapidly beating mouse heart represents perhaps the most challenging context for the use of an optical probe.. . . . the reproducibility and high fidelity of signals obtained. . . provide confidence that genetically encoded sensors will enable the measurement of Ca^{2+} and other cellular signals in virtually any mammalian organ system for which optical access can be obtained, providing a markedly expanded window of observations of physiological and pathophysiological cellular processes." (8, p. 4758)

So far, we have seen how fluorescent proteins can be used to understand the origins of the rhythmic behavior of the heart and the calcium-regulated muscle contractions, but what about the energetics of the heartbeat? Can we use bioluminescence to monitor the power levels of the heart as it is jump-started after a heart attack?

All cells in the human body contain organelles called *mitochondria* that convert the energy contained in food into a more useful form of energy, adenosine triphosphate (ATP), that can be used to power cellular functions. The ATP molecule provides the energy for heat, nerve conduction, and muscle contraction. It is responsible for making our hearts beat and our brains think. ATP is an amazing molecule: right now, each cell in your body contains about 2 billion ATP molecules. As you need energy, the ATP molecules will be broken down. Within 2 minutes, they can all be used up and replaced by new ATP molecules. In a typical day you can produce and use up half your body weight in ATP. (9) ATP is a universal power socket; if something is alive, it has ATP. Every day, the heart creates enough energy in the form of ATP to drive a truck 20 miles. In a human lifetime, that is the equivalent of driving to the moon and back. (10) The mitochondria in the heart muscle cells are clearly busy organelles. Not only do these remarkable power plants produce a lot of ATP, but they are also able to adjust their production levels from resting levels to highly stressed levels when the heart is beating madly.

An average adult mammalian heart muscle cell (*cardiomyocyte*) has approximately 6,000 mitochondria that occupy 40 percent of the cell volume and are rigidly organized. Heping Cheng and his colleagues at the State Key Laboratory of Biomembrane and Membrane Biotechnology in Beijing have used photoactivatable GFP to elucidate intermitochondrial communication in heart muscle cells. Photoactivatable GFP is normally in a dark, nonfluorescent form, but with blue light it can be switched on. By activating photoactivatable GFP in one mitochondria and monitoring all the other mitochondria in the cardiomyocytes, Cheng and his researchers were able to observe that the photoactivatable GFP used two modes of transport to move from mitochondria to mitochondria: a fast transfer between adjacent mitochondria (kissing) and a much slower passage through dynamic nanotubular tunnels (nanotunneling). This led them to propose that "through kissing and nanotunneling, the otherwise static mitochondria in a cardiomyocyte form one dynamically continuous network to share content and transfer signals." (11, p. 2846)

Although we understand how the mitochondria communicate, the exact mechanism by which they regulate the amount of ATP they produce is not fully known. It is a very important function because when the mitochondria don't make enough ATP, the heart doesn't have enough energy to efficiently pump blood around the body, and that can result in a heart attack.

Elinor Griffiths, of the Bristol Heart Institute at the University of Bristol in the United Kingdom, has used a modified virus to introduce the firefly luciferase gene into living heart muscle cells. Because the luciferase lights up only in the presence of ATP, she has created a system to monitor ATP production and breakdown in the heart muscle. Griffiths and her colleagues were particularly interested to discover that in rats there is a lag in ATP production when the heart is restarted after cardiac surgery or a heart attack. Under these conditions, the supply of ATP drops before mitochondrial production starts again, potentially preventing the heart from beating properly. (12, 13)

Oxygen is required to convert the ATP to energy in the cardiomyocytes. Coronary arteries supply the heart muscles with oxygen-laden blood. Blockages of these blood vessels can cause a reduced blood supply to the heart muscles, especially during times of stress and exertion, when the heart needs more oxygen. Diseases of the blood vessels such as cardiovascular disease generally involve narrowed or blocked blood vessels, which can lead to heart attacks or strokes. Chest pains, called *angina*, are the first sign that the heart muscles are deprived of blood. More than 16 million Americans currently have stable angina.

If the flow of blood to a section of the heart muscle is suddenly blocked, that part of the heart muscle is deprived of oxygen, its energy levels begin to fall, and it may cease to contract. This is a heart attack, which is also known as a *myocardial infarction*. If the blockage, which is normally a blood clot, is not removed in a timely manner, oxygen deprivation (*ischemia*) may occur in all cardiac tissue downstream from the obstruction, leading to its cell death (*infarction*). The process is halted when oxygenated blood begins to flow back into the affected areas, which is called *reperfusion*. The amount of damage to the heart depends on the amount of time that has passed between the heart attack and reperfusion. The offending clot is typically removed by a procedure known as an *angioplasty*. Days and even weeks after a heart attack and reoxygenation, previously oxygen-starved myocardial (heart muscle) cells can die from reperfusion injury. This damage may not be obvious, and it may cause severe and long-lasting problems. The mammalian heart has little capacity to regenerate, and the injured myocardial cells are replaced with noncontractile scar tissue.

The goal of cardiovascular regenerative medicine is to replace damaged myocardial cells that have been oxygen deprived due to a heart attack or ischemic heart disease. Many research groups are experimenting with injecting foreign stem cells from blood, bone marrow, or fat tissue into the affected areas of the heart. However, the therapeutic ideal is to stimulate a resident source into replacing the damaged cells. This would overcome the issue of host immune rejection and problems with getting the replacement cells to the site of the injury. Fishes can do this without any prompting; when a fish heart is damaged, it can generate new myocardial cells and fully restore function.

How do fish repair their broken hearts? Prior to 2012, it was only known that adult fish have an ability to regenerate injured heart tissue by proliferation of preexisting heart muscle cells, a talent adult mammals lack. In August 2012, Yasuhiko Kawakami of the Stem Cell Institute at the University of Minnesota and his coworkers reported that the migration of cardiomyocytes to the injury site was critical to the repair process, and that in heart regeneration it was coordinated with the proliferation of cardiomyocytes. (*14*) To visualize the migration of the cardiomyocytes, Kawakami and his colleagues used a unique fluorescent protein named Kaede, which normally fluoresces green but becomes red fluorescent upon exposure to UV light. It was found completely serendipitously. In the paper describing the isolation of the protein and its use, Atsushi Miyawaki wrote: "We happened to leave one of the protein aliquots on the laboratory bench overnight. The next day, we found that the protein sample on the bench had turned red, whereas the others that were kept in a paper box remained green. Although the sky had been partly cloudy, the red sample had been exposed to sunlight through the south-facing windows.. . . To verify this serendipitous observation we put a green sample in a cuvette over a UV illuminator emitting 365-nm light and found that the sample turned red within several minutes.. . . At this point the protein was renamed Kaede, which means maple leaf in Japanese." (*15*, p. 12652)

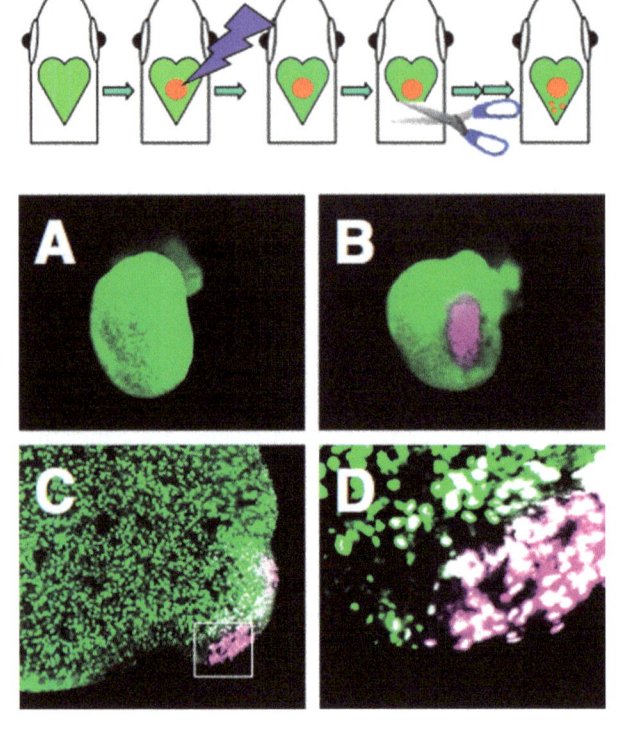

Figure 2.3 Kaede-modified zebrafish heart is green fluorescent when illuminated with visible light (A). Upon exposure to UV light, the cardiomyocytes that were irradiated fluoresce red, while all the others remain green (B). The cardiomyocytes are localized and do not move until the tip of the ventricular portion of the heart was amputated, whereupon they move to the site of the injury and are involved in repair (C and D). (*14*)

Kawakami used zebrafish with Kaede-modified hearts to assay the cardiomyocyte migration. A central spot of the heart was exposed to UV light to convert the Kaede in the illuminated cells from the green into the red form; then the tip of ventricular portion of the heart was amputated. No migration of the red fluorescent Kaede cells was observed in uninjured hearts, while migration to the injury site was observed in the zebrafish with the ventricular apex amputation (figure 2.3).

Like fish, mammalian embryos can repair their own injured hearts. When Hina Chaudhry, associate professor of medicine at Mount Sinai School of Medicine, heard that women in the last few months of pregnancy have the highest rate of recovery from heart failure, she wondered whether it was the fetal cardiomyocytes that were doing the repair. Chaudhry is just the right sort of person to wonder about something like this. She double majored in chemistry and biology at MIT with a thesis in physics and obtained her MD with honors from Harvard Medical School. She has several patents pending for methods to prevent degeneration of heart tissue after heart attack or during heart failure and is the founder and chief scientific officer of a biotech start-up company aimed at developing clinical treatments based on her research findings. In addition to her academic and professional interests, this cardiologist has held leadership positions in several Pakistani-American organizations and has been very involved in supporting the people of Pakistan during the Kashmir earthquake in 2005 and the Sindh flood in 2011.

Using fluorescent proteins, Chaudhry and her co-workers have confirmed their suspicions that fetal stem cells cross the umbilical cord and are attracted to injured maternal hearts, where they are capable of differentiating into a variety of cell types, including beating cardiomyocytes. (16) In experiments very similar to the zebrafish experiments described earlier, researchers in Chaudhry's laboratory used GFP-expressing male mice to impregnate normal virgin female mice to obtain nonfluorescent mice impregnated with fluorescent embryos. Under normal conditions no fluorescent heart cells were observed in pregnant and postnatal females. However, if the heart of the pregnant mother was injured, green fluorescent cardiomyocytes, smooth muscle, and endothelial cells were observed in the injured areas of the heart (see figure 2.4). Thus, the fetus helps to mend its mother's broken heart.

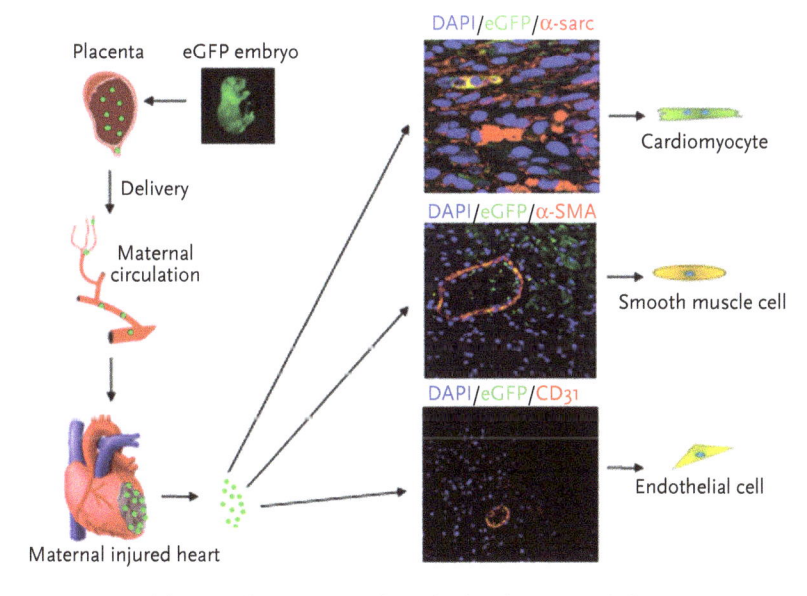

Figure 2.4 Fetal heart cells can move through the placenta to help repair an injured maternal heart. In the experiment illustrated here, the heart of a nonfluorescent mouse has been injured, and the repair of the damage by fluorescent embryonic cells was monitored by fluorescence microscopy. (16)

This is a stunning result. It implies that the heart sends out a chemical distress signal after a heart attack. In all heart attack victims the "Help! Cardiomyocytes in trouble. Please send stem cells" signal echoes futilely throughout the body, except in women in the late stages of pregnancy, where the signal arrives in the embryo, which responds by sending life-saving embryonic stem cells. These stem cells then nearly magically manage to find the source of the chemical distress signal before differentiating into the required heart cells.

Paul Riley, the British Heart Foundation Professor of Regenerative Medicine at Oxford University, and his colleagues from the Molecular Medicine Unit of the Institute of Child Health, University College, London, have shown that stem cells resident in the protective layer of connective tissue covering the heart muscle, the epicardial layer, can be persuaded to form muscle cells within the damaged heart itself. In 2008, Riley was awarded the Outstanding Achievement Award of the European Society of Cardiology Council on Basic Sciences for this finding. In 2011, Riley and his colleagues published a paper in *Nature* in which they described genetically engineered mice with a key embryonic epicardial gene, Wt1, labeled with fluorescent proteins. (17) They found that when a heart attack is induced in their mice, the epicardium of the adult heart reactivates the expression of certain embryonic genes, such as Wt1. If the mice are injected with thymosin β4, which is known to stimulate the epicardium, prior to inducing the heart attack, the Wt1 expressing cells will differentiate into cardiac heart muscle after the heart attack. The newly formed myocardial cells are fully functional and are completely integrated into existing muscle tissue. Pretreatment with thymosin β4 results in better recovery of heart function after infarction. Using the fluorescence of the yellow fluorescent protein (YFP), the authors showed that the Wt1-expressing cells were formed in the epicardium and not in existing muscle cells, and that they subsequently moved and gave rise to a new population of muscle cells. According to Riley and his colleagues "The identification of a bona fide source of myocardial progenitors is a significant step towards resident-cell-based therapy for acute myocardial infarction in human patients." However, they add, "The induced differentiation of the progenitor pool described into cardiomyocytes by thymosin β4 is at present an inefficient process relative to the activated progenitor population as a whole. Consequently, the search is on via chemical and genetic screens to identify efficacious small molecules and other trophic factors to underpin optimal progenitor activation and replacement of destroyed myocardium." (17, p. 644)

Animal model systems are used because it is not practical to study heart muscle cells in live humans, but at some point model systems are no longer sufficient. A solution to this problem would be to study isolated live muscle cells. "Researchers need to study cardiac cells *in vitro* to understand more about the progression of various forms of the disease," David Elliott of Monash University in Australia says. "They need to be able to screen new drugs safely and establish whether or not they are toxic to heart tissue—but obtaining enough mature, live cardiac cells to do this has proved remarkably difficult." (19) Fortunately, that is no longer true. It all changed when Elliott and a team of 26 scientists developed a technique to grow and isolate pure human cardiac cells. Now, if researchers need fully developed cardiomyocytes, they can use Elliott's techniques, or for a price, they can have the cardiomyocytes sent from Australia to their labs anywhere in the world. (18)

"When you grow a culture of embryonic stem cells, you can encourage it to produce specialized cardiac cells using certain growth factors," Elliott says. "But how do you then identify and separate these from the smooth muscle cells and other types in the culture? That was the first big challenge." (19) Elliott's team solved this problem by finding a gene that is activated in the earliest stages of embryonic heart formation. When the gene was tagged with GFP, the stem cells that were about to become cardiac cells glowed green for 3 days before they differentiated into fully developed cardiomyocytes (figure 2.5). The resultant glowing heart cells

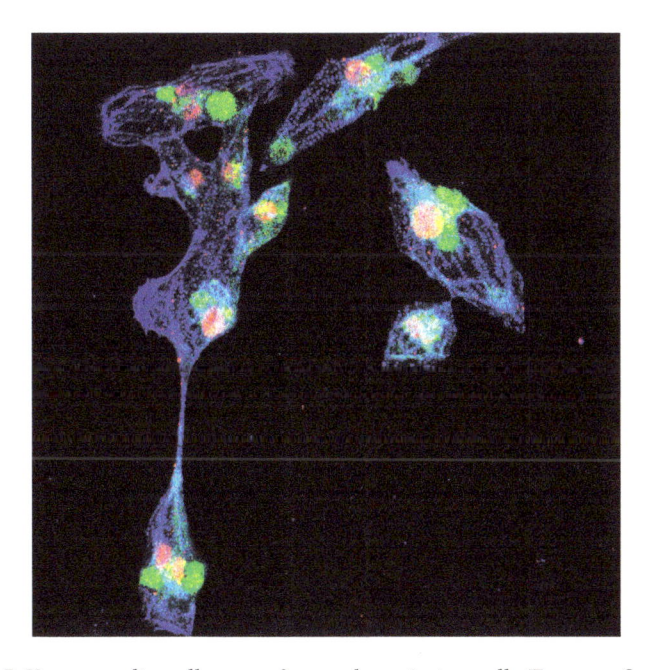

Figure 2.5 Human cardiac cells grown from embryonic stem cells. The green fluorescence has been used to distinguish the cardiomyocytes from other cell types. (18)

were the first pure human cardiac cells, free from contaminating animal products and pathogens. This technique opens the door to personalized cardiomedicine, and biotech companies are already investing in the procedure with the hope that cardiac specialists will be able to isolate heart cells cultured from the stem cells of individual patients. (19)

Ken Suzuki and his coworkers from the National Heart and Lung Institute at Imperial College London have been examining alternative ways of delivering embryonic skeletal muscle stem cells (*myoblast*) for treating rats after they have had a heart attack. To date, most experimental and clinical studies have relied on injecting the skeletal progenitor muscle cells directly into the heart muscles. Suzuki and his team were interested in injecting the skeletal myoblasts into the cardiac veins. They were hoping that cells disseminated in such a gentle manner would be more likely to survive and exhibit an enhanced therapeutic outcome. Five million GFP-expressing skeletal myoblasts were injected directly into the heart muscles or into the cardiac veins of female rats. The injection site had a substantial effect on the distribution of the skeletal myoblasts: those entering through the cardiac veins were widely distributed around the heart and caused no myocardial damage, whereas those delivered right into the heart muscle itself formed clumps and myocardial inflammation. (20)

A heart transplant is the final option for patients with severe coronary artery disease and end-stage heart failure. Christiaan Barnard performed the world's first heart transplant in 1967 in Cape Town, South Africa; the patient died of pneumonia less than a year after the transplant. Since then, the surgical procedure has been improved and the average post–heart transplant survival rate is 15 years, with more than 70 percent of patients undergoing a heart transplant surviving for more than 5 years. Thanks to new alternative treatments and the high price of a heart transplant, about $750,000 in the United States, there is a diminishing demand for heart transplants; nevertheless, there are not nearly enough donors to meet the demand. About 3,500 heart transplants were performed worldwide in 2007, while 800,000 people have heart conditions that warrant a transplant.

There is therefore an urgent need for replacement organs that is not being met by human organ donations, with more than 110,460 people awaiting organ transplants in the United States in 2013. (21) According to the US Food and Drug Administration, every day about 10 people die while on a waiting list for a vital organ transplant. There are two alternatives to heart transplants: the use of artificial hearts or hearts from another species (*xenotransplantation*). (22)

Although primates are our closest relatives, they cannot be bred in captivity in large numbers, and because it is thought that the human immunodeficiency virus (HIV) may have developed from its primate analog, the simian immunodeficiency virus (SIV), there is a fear that primate organs could spread HIV. Pigs are the next best source of organs; their organs are roughly the same size as their human counterparts, and they are evolutionarily distant enough that the two species do not share many pathogens. In order to make porcine transplants a reality, two major obstacles have to be overcome: rejection and disease transmission. Since 2001, many groups have been working-on genetically modified pigs that could be a source of safe transplantation organs that would not be rejected.

Randy Prather, a professor of reproductive biotechnology at the University of Missouri, Columbia, is one of the researchers at the forefront of the creation of transgenic swine for medicine and xenotransplantation. He often uses GFP and its yellow mutant, YFP, as a marker to show that foreign genes can be expressed in transgenic swine. Figure 2.6 shows two pigs. The one on the right is a regular piglet, a little cleaner than a typical piglet, but no different from the piglets that could be found on a hog farm. The one on the left is clearly unlike any pig we have ever seen before. It is a

Figure 2.6 The piglet on the left is a clone with YFP expressed in the tissues of its trotters and snout. The piglet was created in a proof of concept experiment to show that transgenic pigs could be created. Ultimately, the researchers hope to create pigs with human proteins and sugars so that they may become organ donors with porcince body parts that are not detected by the human immunosystem. (Courtesy of University of Missouri Extension and Agriculture Information.)

transgenic cloned pig with enhanced green fluorescent proteins expressed in its snout and trotters. The first genetically modified and cloned pig ever produced, it was created by Prather's lab in 2001 to show that it is possible to produce a transgenic clone. According to Prather and his colleagues, "These animals prove that we can make genetic modifications to express desired traits. For xenotransplantation, this is a large step because it means it's possible to change the genetic makeup of the cells to prevent the body's rejection of transplanted organs." (23, p. 213)

The piglet was cloned using a process called *nuclear transfer*, the same method that was used to clone Dolly the sheep. But before the piglet could be cloned, the GFP gene had to be inserted into its genome. This was done by removing some cells from a pig, growing them in the laboratory, and adding the GFP gene to the genes responsible for producing the tissues in the hooves and snout. Now that the DNA in the nuclei of these cells contained all the information required to make an identical copy of the original pig, with the addition of some genetically altered GFP, they were removed from the cells by using a thin, hollow needle. Then the modified DNA was inserted into the nucleus of mature unfertilized eggs, called *oocytes*, from which the DNA had been removed. Five such "fertilized" eggs were implanted into the uterus of a surrogate mother, which carried the piglets until birth. Four of the resulting piglets had enhanced green fluorescent proteins in their hooves and snout (see figure 2.6). They were all identical to the pig that donated its cells at the beginning of the experiment, but they had a mutated jellyfish gene expressing enhanced GFP in their snouts and trotters. (24)

At this point you might be asking yourself why Prather created a transgenic clone and not just a transgenic pig. What is the advantage of using a cloned pig? The answer is quite simple: by using clones of known healthy pigs, one can be assured that their transgenically modified "offspring" will be free of genetic diseases.

It has been more than 10 years since the first transgenic cloned pig was created. How close are we to being able to transplant major organs from pigs to humans?

Currently, genetically engineered pig hearts placed in baboons have functioned for periods of 3 to 6 months, kidneys for periods approaching 3 months, livers for a matter of days, and lungs for a matter of hours. Although porcine heart xenotransplantation has moved past the first step to implementation and issues of immunorejection, and potential infections by pig endogenous retroviruses have been reduced, new problems such as mini-clots forming in capillaries have emerged. Despite the slow progress and numerous detractors, those working in the field are optimistic.

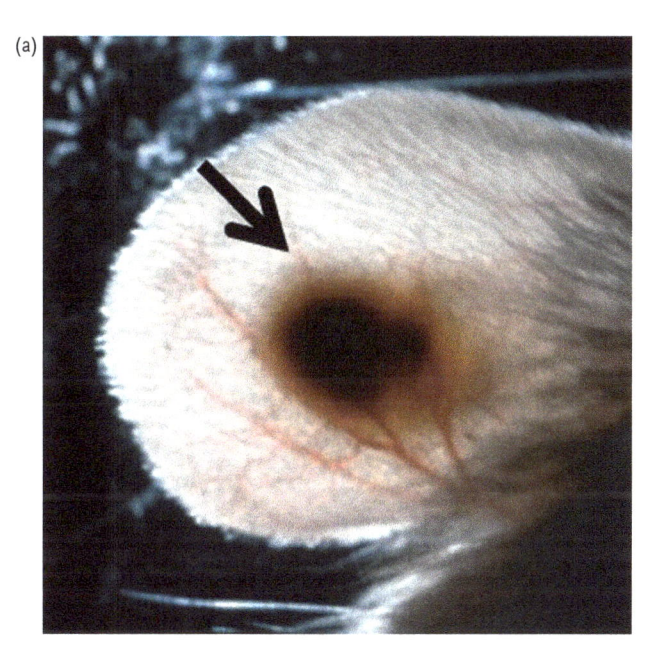

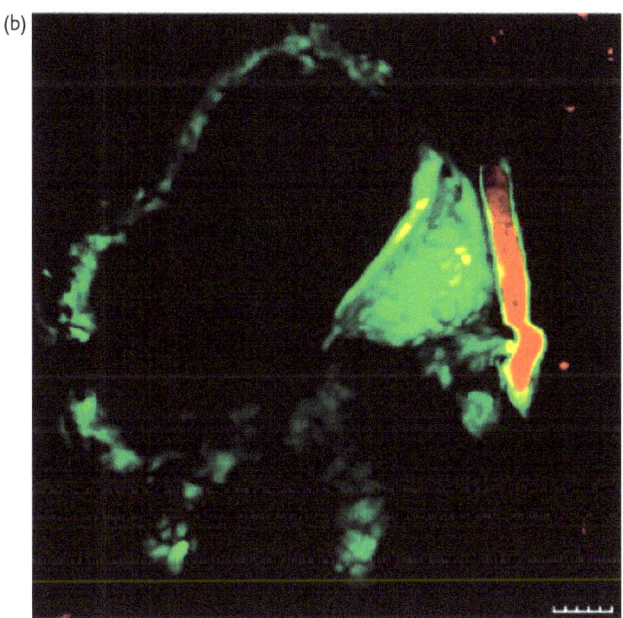

Figure 2.7 Fully functional heart beating in the earlobe of a mouse (a). Fluorescent image of GFP heart 9 weeks after transplantation (b). The blood vessels are highlighted with a red dye. (*26*)

For example, Brucin Ekser and David Cooper from the Transplantation Institute at the University of Pittsburgh Medical Center say, "The potential therapeutic possibilities offered by xenotransplantation are so considerable that it remains an area of research that should be pursued vigorously until the barriers have been overcome. Not only will pig organs and islets (for diabetes) offer therapeutic options, but there are potential therapies related to pig corneal transplants, pig neural-cell transplants (in conditions such as Parkinson's and Huntington's disease), and even pig red blood cells for transfusion into humans. The number of patients who might benefit from xenotransplantation may therefore run into the hundreds of thousands or even millions if it can achieve its potential." (25, p. 226)

It is extremely difficult to track the fate of individual cardiac cells in a living mouse. They are just too deep inside the animal, and consequently the light has to penetrate the flesh and avoid bony structures. Benny Chen from the Duke University Medical Center has devised a unique workaround to this problem. He and his team have implanted a fluorescent heart from a newborn mouse under the skin of a mouse earlobe (figure 2.7). It may sound crazy, but it works. Because the heart is close to the surface and the skin of the ear is much thinner than elsewhere, the three-dimensional wanderings of individual heart cells can be tracked by two-photon microscopy. And, perhaps most important, the heart in the ear is a good model for the real thing. According to Chen and his colleagues, "All hearts will begin to beat if the surgery is successful. If the heart is accepted it can survive and beat indefinitely; whereas if the heart is rejected, it will stop beating after about 10 days." This is not surprising because "investigators in the field of transplantation have been using an 'ear-heart' murine model to study immune tolerance for more than 40 years." (26, p. e52087)

I began this chapter on heart disease by asking, "How is that reassuring rhythm of a parent's or lover's heartbeat regulated?" and proceeded to describe the work that illuminated a rhythm protein that directs the symphony of life and is responsible for the comforting thumps we can hear emanating from a loved one's chest. At this point, I have to wonder how irritating that same heartbeat must be if the heart is located inside your earlobe.

REFERENCES

1. Centers for Disease Control, Faststats, Heart Disease. (2013). www.cdc.gov/nchs/fastats/heart.htm.
2. Wade, N. (1998). Animal's genetic program decoded, in a science first. *New York Times*, December 11, A1.

3. Norman, K. R., Fazzio, R. T., Mellem, J. E., Espelt, M. V., Strange, K., Beckerle, M. C., and Maricq, A. V. (2005). The Rho/Rac-family guanine nucleotide exchange factor VAV-1 regulates rhythmic behaviors in C-elegans. *Cell 123*, 119–132.
4. University of Utah. (2005). Rhythm gene discovered: The scoop on when worms poop, ovulate and swallow. Press release. www.sciencedaily.com/releases/2005/10/051007094136.htm.
5. Sauzeau, V., Jerkic, M., Lopez-Novoa, J. M., and Bustelo, X. R. (2007). Loss of Vav2 proto-oncogene causes tachycardia and cardiovascular disease in mice. *Molecular Biology of the Cell 18*, 943–952.
6. Tsiaras, A. (2005) *The InVision guide to a healthy heart*. New York: HarperCollins.
7. Ramanujan, K. (2005). Mice with glowing hearts shed light on living cells. Press release, Cornell University news service.
8. Tallini, Y. N., Ohkura, M., Choi, B.-R., Ji, G., Imoto, K., Doran, R., Lee, J., Plan, P., Wilson, J., Xin, H.-B., Sanbe, A., Gulick, J., Mathai, J., Robbins, J., Salama, G., Nakai, J., and Kotlikoff, M. I. (2006). Imaging cellular signals in the heart in vivo: Cardiac expression of the high-signal Ca2+ indicator GCaMP2. *Proceedings of the National Academy of Sciences of the United States of America 103*, 4753–4758.
9. Bryson, B. (2003). *A short history of everything*. New York: Broadway Books.
10. Stein, L., Gura, T., Thompson, J. M., Daniels, P., Hitchcock, S. T., Bechtel, S., and Restak, R. M. (2007). *Body: The complete human*. Washington, DC: National Geographic.
11. Huang, X. H., Sun, L., Ji, S. X., Zhao, T., Zhang, W. R., Xu, J. J., Zhang, J., Wang, Y. R., Wang, X. H., Franzini-Armstrong, C., Zheng, M., and Cheng, H. P. (2013). Kissing and nanotunneling mediate intermitochondrial communication in the heart. *Proceedings of the National Academy of Sciences of the United States of America 110*, 2846–2851.
12. Griffiths, E. (2006). Luciferase throws light on heart disease. TrAC *Trends in Analytical Chemistry 25*, III.
13. Bell, C. J., Bright, N. A., Rutter, G. A., and Griffiths, E. J. (2006). ATP regulation in adult rat cardiomyocytes: Time-resolved decoding of rapid mitochondrial calcium spiking imaged with targeted photoproteins. *Journal of Biological Chemistry 281*, 28058–28067.
14. Itou, J., Oishi, I., Kawakami, H., Glass, T. J., Richter, J., Johnson, A., Lund, T. C., and Kawakami, Y. (2012). Migration of cardiomyocytes is essential for heart regeneration in zebrafish. *Development 139*, 4133–4142.
15. Ando, R., Hama, H., Yamamoto-Hino, M., Mizuno, H., and Miyawaki, A. (2002). An optical marker based on the UV-induced green-to-red photoconversion of a fluorescent protein. *Proceedings of the National Academy of Sciences of the United States of America 99*, 12651–12656.
16. Kara, R. J., Bolli, P., Karakikes, I., Matsunaga, I., Tripodi, J., Tanweer, O., Altman, P., Shachter, N. S., Nakano, A., Najfeld, V., and Chaudhry, H. W. (2011). Fetal cells traffic to injured maternal myocardium and undergo cardiac differentiation. *Circulation Research 110*, 82–93.
17. Smart, N., Bollini, S., Dube, K. N., Vieira, J. M., Zhou, B., Davidson, S., Yellon, D., Riegler, J., Price, A. N., Lythgoe, M. F., Pu, W. T., and Riley, P. R. (2011). De novo cardiomyocytes from within the activated adult heart after injury. *Nature 474*, 640–644.
18. Elliott, D. A., Braam, S. R., Koutsis, K., Ng, E. S., Jenny, R., Lagerqvist, E. L., Biben, C., Hatzistavrou, T., Hirst, C. E., Yu, Q. C., Skelton, R. J. P., Oostwaard, D. W. V., Lim, S. M., Khammy, O., Li, X. L., Hawes, S. M., Davis, R. P., Goulburn,

A. L., Passier, R., Prall, O. W. J., Haynes, J. M., Pouton, C. W., Kaye, D. M., Mummery, C. L., Elefanty, A. G., and Stanley, E. G. (2011). NKX2-5eGFPw hESCs for isolation of human cardiac progenitors and cardiomyocytes. *Nature Methods* 8, 1037–1040.
19. Cribb, J., and Roginski, A. (2012). Glowing cells to help mend broken hearts. *Monash University Magazine*, October. http://www.monash.edu.au/monashmag/articles/glowing-cells-to-help-mend-broken-hearts.html#.U53tXSjm_y8
20. Fukushima, S., Coppen, S. R., Lee, J., Yamahara, K., Felkin, L. E., Terracciano, C. M. N., Barton, P. J. R., Yacoub, M. H., and Suzuki, K. (2008). Choice of cell-delivery route for skeletal myoblast transplantation for treating post-infarction chronic heart failure in rat. *PLOS One* 3(8), e3071.
21. United Network for Organ Sharing. http://optn.transplant.hrsa.gov/. Accessed January 4, 2013.
22. US Food and Drug Administration. Vaccines, blood and biologics—Xenotransplantation. www.fda.gov/BiologicsBloodVaccines/Xenotransplantation/default.htm.
23. Cabot, R. A., Kuhholzer, B., Chan, A. W. S., Lai, L., Park, K. W., Chong, K. Y., Schatten, G., Murphy, C. N., Abeydeera, L. R., Day, B. N., and Prather, R. S. (2001). Transgenic pigs produced using in vitro matured oocytes infected with a retroviral vector. *Animal Biotechnology* 12, 205–214.
24. Park, K. W., Cheong, H. T., Lai, L. X., Im, G. S., Kuhholzer, B., Bonk, A., Samuel, M., Rieke, A., Day, B. N., Murphy, C. N., Carter, D. B., and Prather, R. S. (2001). Production of nuclear transfer–derived swine that express the enhanced green fluorescent protein. *Animal Biotechnology* 12, 173–181.
25. Ekser, B., and Cooper, D. K. C. (2010). Overcoming the barriers to xenotransplantation: Prospects for the future. *Expert Review of Clinical Immunology* 6, 219–230.
26. Chen, B. J., Jiao, Y. Q., Zhang, P., Sun, A. Y., Pitt, G. S., Deoliveira, D., Drago, N., Ye, T., Liu, C., and Chao, N. J. (2013). Long-term in vivo imaging of multiple organs at the single cell level. *PLOS One* 8(1), e52087.

CHAPTER 3

Chagas

Worldwide (infected)	10–18 million
United States (infected)	300,000
Develop chronic cardiac problems	30%
Develop chronic gastric problems	10%

Endemic Chagas disease has emerged as an important health disparity in the Americas. As a result, we face a situation in both Latin America and the US that bears a resemblance to the early years of the HIV/AIDS pandemic.

Editorial, *PLOS Neglected Disease*, May 2012

Chagas disease, aka the kissing disease, or American trypanosomiasis, is a disorder with three names that kills more people than any other parasitic illness in South America, and yet it is one of the least known diseases described in this book. In some areas of South America, at least 1 in 16 people are infected with Chagas. It has existed in wild animals for more than 10 million years, and analyses of Peruvian and Chilean mummies have revealed that the disease has been infecting residents of South America for more than 9,000 years.

Bloodsucking triatomine bugs, nicknamed "kissing bugs," carry the parasite *Trypanosoma cruzi*, that is responsible for Chagas disease. These bugs get their moniker because they often bite their victims in the face while they are sleeping. During or immediately after biting their victims, the kissing bugs defecate near the bite wound. I am not sure where they learned their manners or what their mothers think of this behavior. The infective form of the *T. cruzi* parasite is in the feces. The kissing bug bites are itchy, but scratching is a bad idea because it allows the parasites to enter their new host when the scratching spreads the feces over the wound. Kissing bugs often bite near their victim's eyes. Smearing the feces into the eyes provides another way for the parasite to enter the host, and a swelling of the eye that is often associated with Chagas disease occurs.

If the kissing bug bites a victim who is infected with *T. cruzi*, the parasite enters the insect during its blood meal. Once inside the insect, *T. cruzi* multiplies in the midgut before being transformed into a new infective form in the rectum that is ready to start a new cycle. The parasite undergoes a complex lifecycle and can be found in four distinct morphological forms, two of which are replicative stages.

Most of the world's Chagas research is conducted in Brazil, with the Instituto Carlos Chagas in Curitiba one of the main Chagas research institutes. There, Stenio Fragoso and his colleagues have developed transgenic *T. cruzi* parasites that express GFP only when they are in their reproductive life forms. The genetically modified parasites can be used to infect cultured mammalian cells. And at any stage the cells can be separated, sorted by fluorescence, and counted using fluorescence cytometry, a technique we will encounter later in the chapter on malaria. Because the fluorescence is limited to *T. cruzi* in the replicative life forms, the transgenic GFP parasites are useful in studying gene expression regulation, host-parasite interaction, drug screening, and vaccine development against the replicative stage. (1)

The disease can also be spread by blood transfusions and in food contaminated with the parasite. In Brazil, insect feces in homemade fruit juices have been shown to be a significant vector for the spread of the disease. The triatomine bugs are not particularly fussy and are known to feed on the blood of more than 150 species of mammals. All these species can host the *T. cruzi* parasite. While the kissing bug can transmit the parasite to humans only through fecal transmission, in other lower-order mammals the Chagas infection can be spread by eating whole infected triatomine bugs. Although the kissing bugs don't look like a fetching meal, they resemble cockroaches that can grow an inch or two in length; numerous smaller mammals enjoy dining on them.

Kissing bugs hide out in dark, protected places like cracks in the walls of poorly constructed homes and thatch roofs. More than 14,000 kissing bugs can be found in a single chronically infested home. (2) At night they leave their hidey-holes and follow the plumes of carbon dioxide exhaled by their sleeping victims. Once they have found their victims, the kissing bugs unfold their hair-like proboscises and have a leisurely 5- to 10-minute blood meal.

An old and understandably unpopular, but still commonly used, test for Chagas disease is a xenodiagnosis. In this test, parasite-free kissing bugs are placed on the patient and are allowed to have an undisturbed blood banquet. Sixty days later the feces of the kissing bugs are examined for trypanosomes.

It is estimated that globally between 10 and 18 million people have Chagas disease and approximately one-third of them will die from heart and digestive failure directly related to this illness. These numbers cover such a large range because Chagas disease is often misdiagnosed. At first the disease masquerades as the flu, with body aches and a fever that may last for a few weeks. If the parasitic infection was spread by a bug bite, a swelling at the site of the bite can be an indicator for the disease. These symptoms will disappear, but the parasite remains in its host, where it can live for decades before reappearing to cause digestive and heart disorders. These symptoms are often falsely associated with aging and not Chagas disease. During the dormant phase, few if any parasites are found in the patient's blood, and most people are unaware of the infection. Approximately 20 to 30 percent of infected individuals will progress to a chronic stage, where they experience heart rhythm abnormalities that can cause heart failure and strokes in patients as young as 30. The average life expectancy of someone with chronic cardiac Chagas is about 10 years. "Your heart just turns into a big, ineffective bag," said cardiologist Sheba Meymandi, the director of a Chagas treatment program at the UCLA Medical Center, the only hospital with a program devoted to treating patients with Chagas disease in the United States. About 10 percent of the parasite's victims progress to the chronic gastric form of the disease, which causes a dilated esophagus or colon, leading to difficulties with eating and passing stool. Fortunately, the other 70 to 80 percent never develop any further Chagas-related symptoms.

Part of the reason Chagas disease is often misdiagnosed or not diagnosed at all is because it is a neglected disease often associated with poverty and highly stigmatized. After hookworm, it is the most common disease afflicting the estimated 99 million people who live on less than two dollars a day in the Latin American and Caribbean region. (3) There is no vaccine against Chagas disease, and most preventive measures rely on controlling the number of triatomine bugs with pesticides.

Currently, two drugs are available to treat Chagas, nifurtimox and benznidazole, but both have severe side effects and are therefore unsatisfactory. Early treatment is life-saving but expensive, with a lifetime cost averaging $11,619 per patient. In 2008, no more than $15.6 million was spent globally on Chagas research, and less than $5 million of that money was used for finding new diagnostics, drugs, and vaccines. (4) Chagas patients cannot afford expensive new treatment, and consequently there is no financial incentive for pharmaceutical companies to develop new and improved drugs against *T. cruzi*. In contrast, thanks to a vocal and committed activist community, HIV/AIDS research has received a tremendous amount of publicity, and widespread access to antiretroviral treatments is available in developed and developing countries. Benznidazole is the only Chagas therapy used in Latin America, and acute shortages of this drug are postponing the treatment for thousands of Chagas patients. (3) According to Peter Hotez and his co-workers at the Sabin Vaccine Institute and Texas Children's Center for Vaccine Development at Baylor College of Medicine, "[Chagas disease] translates into a humanitarian catastrophe for the poorest people in the Americas and elsewhere. This perceptible health disparity demands urgent attention by global health policy makers to prioritize Chagas disease and develop a comprehensive strategy for control and elimination efforts, blood screening and point-of-care testing, maternal and child interventions, health education, and parallel research and development." (3, p. e1498)

The most effective way of controlling the spread of Chagas disease is to get rid of kissing bug infestations in the places where people sleep. A major international control program in Argentina, Bolivia, Brazil, Chile, Paraguay, and Uruguay initiated in 1991 has been very successful. Uruguay was certified free of vector-borne transmission in 1997, Chile followed in 1999, and Brazil was declared transmission-free in 2006. Similar programs are being instituted in Central America and Mexico, where Chagas disease is still prevalent.

Carlos Justiniano Ribeiro Chagas, a young physician born in Brazil on July 9 1879, was the first and probably only person to discover and completely describe a previously unknown disease. Although his parents were well-off and owned a small coffee farm, Chagas's life was not easy. His father died when he was just 4 years old, and his mother, who was only 24 when her husband passed away, raised Carlos. At 14 he was sent to a school for mining engineering. However, Carlos Chagas was not cut out to be an engineer, and he switched to medicine when he was 20. He wrote his MD thesis on malaria and got a job working on malaria control with the Brazilian government. In 1909 it posted him to Lassance, a very small town 350 miles north of Rio de Janeiro, where he was entrusted with the impossible task of ensuring that workers who were trying to lay railway tracks up to the Amazon did not contract malaria. In a railroad car that doubled as his home and clinic, he discovered that the workers were suffering from an unknown disease. Chagas described the parasite responsible for the disease and named it *Schizotrypanum cruzi* (it was later renamed *Trypanosoma cruzi*). He reported that parasitical infections were caused by triatomine bug bites, listed the clinical consequences of the disease, and described its epidemiology. His only mistake was his belief that it was the bite of the kissing bug that transferred the parasite, which is not surprising because that would be a much more efficient way of infecting a victim than depositing the parasite in feces and waiting for them to be smeared across the bite or into the mucosa of the eye. Chagas initially found the parasite in the blood of a 3-year-old. By the time Chagas died of a heart attack at 55 years of age, he had described more than 100 cases of the chronic form of the disease and 27 cases of the acute disease. Chagas was the first Brazilian to be awarded an honorary doctorate by Harvard University. He was nominated for the Nobel Prize in 1913 and 1928 but, controversially, was never awarded the honor. (5)

Ravi Durvasula of the Center for Global Health in the Department of Internal Medicine at the University of New Mexico and his collaborators have been using biomedical techniques to combat Chagas. They have genetically modified bacteria that live in the guts of kissing bugs so that the bacteria produce proteins that punch holes in the walls of the Chagas parasite, as well as proteins that flush all the parasites out of the kissing bug.

To get the transgenic bacteria in the kissing bug's guts, Durvasula and his co-workers have made an artificial feces-like paste that is filled with the modified bacteria. Infant kissing bugs love to wallow in adult fecal matter; it's an evolutionary mechanism designed to fill the growing bug with nutrient-producing bacteria found in the adult dung. "The paste is really fake bug feces," Durvasula says. "But that doesn't sound so good, so we call it Cruzigard." Experiments have shown that the dung-wallowing bacterial transfer to the juvenile kissing bugs in Cruzigard works. (6, p. 94)

Although the genetically modified bacteria have proved to be very effective, Ravi Durvasula is concerned that they will develop resistance to the two proteins his bacteria are releasing in the kissing bug's digestive system. That is why he is hunting for more *T. cruzi*–targeting proteins that he can introduce into the bugs' guts with his genetically modified bacteria. "The idea is to create a kind of genetic factory of different mechanisms," Durvasula explains. "Like beads on a necklace, we can shift from one to the next in order to anticipate mutations." (6, p. 95)

The newest weapons in Durvasula's transgenic bacteria's arsenal are red fluorescent proteins that are flanked by two molecules that bind tightly to sugars on the surface of *T. cruzi*. Durvasula and his fellow researchers call this red fluorescent sugar binder REDantibody. Sugars on the surface of

T. cruzi have to bind molecules that are sticking out of the kissing bug's gut lining in order for the Chagas parasite to mature into an infectious form inside the insect. REDantibody attacks these sugars. Figure 3.1 shows a *T. cruzi* engulfed in the REDantibodies. Not only does the red fluorescence of the modified red fluorescent protein act as an excellent tag to follow the parasite, but, more important, it coats the surface of *T. cruzi*, thereby blocking all of its binding sites from docking with the gut lining of the kissing bug. (7, 8) It is the third weapon added to the genetically modified anti–*T. cruzi* bacteria.

Durvasula and his coworkers have built model Guatemalan thatch huts located in triply protected level 2 biological hazard greenhouses to test the effectiveness of their transgenic anti–*T. cruzi* bacteria. Fake feces full of transgenic bacteria, which were smeared in the thatch hut, proved irresistible to the young kissing bugs. They romped around in the bacteria-laden faux poop, giving the bacteria access to the bug's stomachs. There the bacteria produced their three foreign anti–*T. cruzi* proteins, which floated around harmlessly in the kissing bug gastric juices until the kissing bug was infected by the *T. cruzi*. Then the proteins swung into action. One of the proteins punched a hole in the walls of the Chagas parasite, the other coated it in fluorescent proteins, and the third flushed them out before they evolved into their infective forms. The tests worked well, and preparations are underway for carefully controlled field tests. (8, 9)

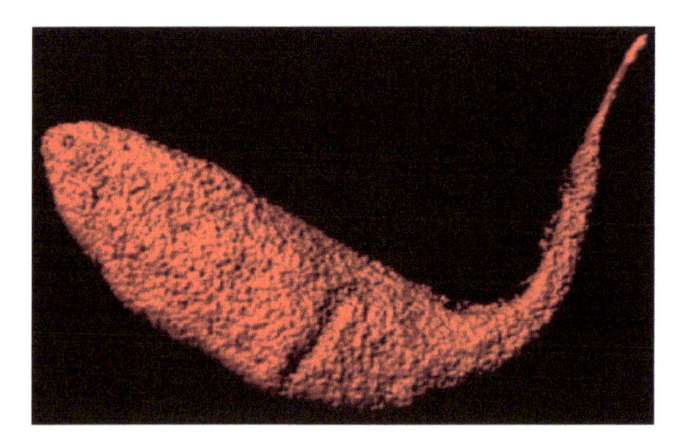

Figure 3.1 This confocal microscopy image shows a Chagas parasite, *T. cruzi*, engulfed by red fluorescent proteins designed to bind with sugar molecules on the surface of the parasite. The red fluorescence allows scientists to track the parasite. More important, the binding of the modified red fluorescent proteins to the surface sugars blocks the parasite from binding to the lumen of the kissing bug and prevents it from reaching its infective life stage. (7)

Have you ever taken a flashlight and held your hand over its light? If so, you will have noticed that all the wavelengths except red were absorbed, and you could see only red light. Before 1999, no red fluorescent proteins were known, and the use of fluorescent proteins as imaging agents in vivo was limited by the poor penetration of the excitation light. Because hemoglobin, water, and lipids minimally absorb red and infrared wavelengths, imaging agents that are red fluorescent were sought. Groups all over the world tried mutating GFP to form a red GFP mutant. This strategy was not very successful, and we would have to wait until 2008 before a red mutant of *Aequorea victoria* GFP was created. (*10*) Other groups took to oceans to look for red bioluminescent organisms (organisms that were giving off red light). They were no more successful.

Mikhail Matz, a graduate student working with Sergey Lukyanov at the Russian Academy of Sciences, was also unsuccessful in his search for organisms that give off red light—that is, until Matz had lunch with Yulii Labas, an expert in the evolution of bioluminescent systems at the Institute of Ecology and Evolution of the Russian Academy of Sciences. Labas pointed out that many of the species that Matz and Lukyanov were planning on examining had recruited their bioluminescent proteins from various biochemical pathways that had nothing to do with light production. Maybe GFP was also a "recent recruit," and perhaps some of its analogs could be found in species that fluoresce but do not bioluminesce. "For instance a friend of mine has a reef aquarium with many representatives of the Anthozoa class, and although none of these species posses bioluminescent capabilities, they are very green-fluorescent! Why not free your minds and try these?" Labas said. (*11*, p. 954) It took a conceptual shift to find red fluorescent proteins. Thinking that aequorin and GFP might have evolved separately and that fluorescent proteins did not necessarily have to be associated with other chemiluminescent proteins, Matz and Lukyanov looked for organisms that were red fluorescent but were not bioluminescent. (*12*) In aquarium shops in Moscow, they found corals containing the first "red" fluorescent protein, DsRed. According to Sergey Lukyanov, "In comparison to many other marine coelenterates, working on reef *Anthozoa* is particularly convenient since one can readily buy the specimens for several dollars in aquarium shops throughout the world. We did not organize expeditions to collect our first samples, but found brightly fluorescent and colored sea anemones and corallimorphs (mushroom anemones) in Moscow instead." Since Lukyanov and Matz found DsRed in 1999, more than 150 distinct fluorescent or colored GFP-like proteins have been reported. In August 2007, Lukyanov reported a bright, fast-folding fluorescent protein that emits light in the far-red part of the spectrum. The protein is named Katushka, a diminutive form of Ekaterin; it is named after Ekatrina Merzlyak, one of the researchers working with Lukyanov, and was isolated from a brilliant red sea anemone that was also bought in a Moscow pet shop. (*13, 14*) Katushka is much brighter than the DsRed (figure 3.2).

Now we know why Durvasula and many others working with fluorescent proteins prefer using red fluorescent proteins over GFP. The fluorescence of the red sugar binders enveloping the Chagas parasites (see figure 3.1), can be observed deeper in infected mice than one would be able to see an equivalent green fluorescent parasite.

Trypanosoma rangeli is a very close relative of *T. cruzi*; both are protozoan parasites that inhabit the same insect vectors and mammalian hosts, but in some ways *T. rangeli* acts exactly the opposite to *T. cruzi*. It prefers taking the direct route between hosts and is transmitted by the

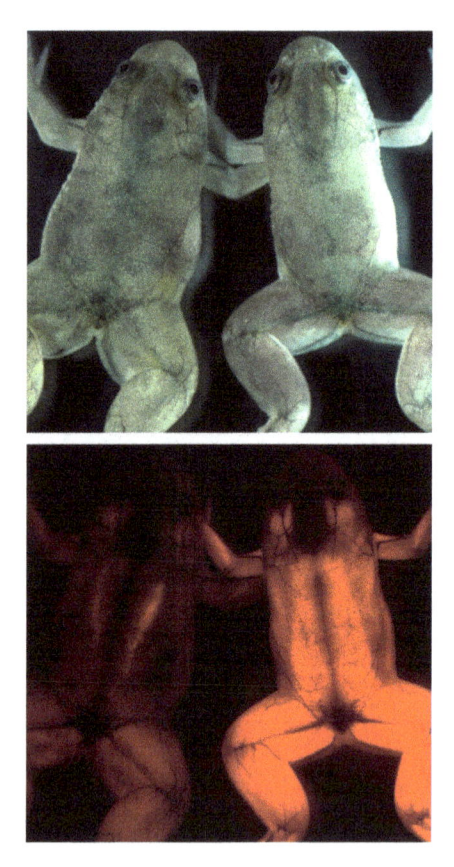

Figure 3.2 Transgenic 2.5-month-old frogs expressing Katushka (right) and DsRed-Express (left). The frogs are shown from the back side under white light. (*14*) (Photo courtesy of Andrey Zaraisky.)

bug bite; there is no fecal ride for *T. rangeli*. Infections of *T. rangeli* don't cause Chagas disease and are harmless to their human hosts. However, the kissing bug needs to be on guard because *T. rangeli* has pathological effects on the insect. It is fairly common to find both parasites present in their mammalian and triatomine hosts.

Palmira Guevara and her coworkers at the Institute of Experimental Biology at the University of Central Venezuela have genetically modified both parasites so that *T. rangeli* expresses GFP and *T. cruzi* expresses the red fluorescent protein, DsRed. She infected some triatomine bugs by feeding them with blood containing 2,000 fluorescent parasites per milliliter. Thanks to the fluorescent tags, the parasites could be followed and distinguished from each other anywhere in the insect. (15)

In kissing bugs infected with both parasites, Guevara noticed that the red fluorescent *T. cruzi* remain in the insect's gut, where they transform into the infective form and exit via the anus. At the same time, large numbers of the green fluorescent *T. rangeli* cross the intestinal lining and enter the insect's bloodstream. This explains why *T. rangeli* can directly enter the bloodstream during a blood meal, while *T. cruzi* has to use the more inefficient route and has to enter its human hosts via infected feces.

One of the treatment options for patients with end-stage Chagas heart disease is a heart transplant. The long-term survival rate of Chagas patients with heart transplants is significantly better than that of patients who have had transplants for other reasons, probably because *T. cruzi* infections are often limited to the heart. (16) However, heart transplants are very expensive, and no more than a hundred Chagas-related transplants have been performed in the United States.

To eliminate the need for heart transplants due to Chagas, it is necessary to attack the Chagas parasite and its insect hosts on all fronts. Many South American countries have vanquished Chagas disease from their borders, with new pesticides and transgenic bacteria, and with a vocal and activist community it should be possible to rid the remaining South American and Central American countries of new Chagas infections. If Chagas disease becomes as well known as HIV/AIDS, or if it becomes a cause célèbre like malaria, it will be a disease whose days are numbered.

REFERENCES

1. Kessler, R. L., Gradia, D. F., Rampazzo, R. D. P., Lourenco, E. E., Fidencio, N. J., Manhaes, L., Probst, C. M., Avila, A. R., and Fragoso, S. P. (2013). Stage-regulated GFP expression in Trypanosoma cruzi: Applications from host-parasite interactions to drug screening. *PLOS One* 8(6), e67441.

2. Coura, J. R., and Vinas, P. A. (2010). Chagas disease: A new worldwide challenge. *Nature 465*, S6–S7.
3. Hotez, P. J., Dumonteil, E., Woc-Colburn, L., Serpa, J. A., Bezek, S., Edwards, M. S., Hallmark, C. J., Musselwhite, L. W., Flink, B. J., and Bottazzi, M. E. (2012). Chagas disease: The new HIV/AIDS of the Americas. *PLOS Neglected Tropical Disease 6*, e1498.
4. Clayton, J. (2010). Chagas disease 101. *Nature 465*, S4–S5.
5. Krasner, R. I. (2008). *20th century microbe hunters: Their lives, accomplishments, and legacies*. Sudbury, MA: Jones and Bartlett.
6. Hitt, J. (2001). Case study: Chagas disease: Location: Guatemala: Building a better bloodsucker. *New York Times Magazine*, May 6, 92–96.
7. Markiv, A., Anani, B., Durvasula, R. V., and Kang, A. S. (2011). Module based antibody engineering: A novel synthetic REDantibody. *Journal of Immunological Methods 364*, 40–49.
8. Hurwitz, I., Fieck, A., Read, A., Hillesland, H., Klein, N., Kang, A., and Durvasula, R. (2011). Paratransgenic control of vector borne diseases. *International Journal of Biological Sciences 7*, 1334–1344.
9. Durvasula, R. V., Kroger, A., Goodwin, M., Panackal, A., Kruglov, O., Taneja, J., Gumbs, A., Richards, F. F., Beard, C. B., and Cordon-Rosales, C. (1999). Strategy for introduction of foreign genes into field populations of Chagas disease vectors. *Annals of the Entomological Society of America 92*, 937–943.
10. Mishin, A. S., Subach, F. V., Yampolsky, I. V., King, W., Lukyanov, K. A., and Verkhusha, V. V. (2008). The first mutant of the Aequorea victoria green fluorescent protein that forms a red chromophore. *Biochemistry 47*, 4666–4673.
11. Matz, M. V., Lukyanov, K. A., and Lukyanov, S. A. (2002). Family of the green fluorescent protein: Journey to the end of the rainbow. *Bioessays 24*, 953–959.
12. Matz, M. V., Fradkov, A. F., Labas, Y. A., Savitisky, A. P., Zaraisky, A. G., Markelov, M. L., and Lukyanov, S. A. (1999). Fluorescent proteins from nonbioluminescent Anthozoa species. *Nature Biotechnology 17*, 969–973.
13. Merzlyak, E. M., Goedhart, J., Shcherbo, D., Bulina, M. E., Shcheglov, A. S., Fradkov, A. F., Gaintzeva, A., Lukyanov, K. A., Lukyanov, S., Gadella, T. W. J., and Chudakov, D. M. (2007). Bright monomeric red fluorescent protein with an extended fluorescence lifetime. *Nature Methods 4*, 555–557.
14. Shcherbo, D., Merzlyak, E. M., Chepurnykh, T. V., Fradkov, A. F., Ermakova, G. V., Solovieva, E. A., Lukyanov, K. A., Bogdanova, E. A., Zaraisky, A. G., Lukyanov, S., and Chudakov, D. M. (2007). Bright far-red fluorescent protein for whole-body imaging. *Nature Methods 4*, 741–746.
15. Guevara, P., Dias, M., Rojas, A., Crisante, G., Abreu-Blanco, M. T., Umezara, E., Vazquez, M., Levin, M., Anez, N., and Ramirez, J. L. (2005). Expression of fluorescent genes in Trypanosoma cruzi and Trypanosoma rangeli (Kinetoplastida: Trypanosomatidae): Its application to parasite-vector biology. *Journal of Medical Entomology 42*, 48–56.
16. Bocchi, E. A., and Fiorelli, A. (2001). The paradox of survival results after heart transplantation for cardiomyopathy caused by Trypanosoma cruzi. *The Annals of Thoracic Surgery 71*, 1833–1838.

CHAPTER 4

Malaria

Malaria Fatalities, 2010

Worldwide	655,000
Africa	596,000
America	1,000
Children under 5 years of age	563,000

Their bite is as painful as that of the serpents, and causes diseases.... [The wound] as if burnt with caustic or fire, is red, yellow, white, and pink color, accompanied by fever, pain of limbs, hair standing on end, pains, vomiting, diarrhea, thirst, heat, giddiness, yawning, shivering, hiccups, burning sensation, intense cold.

Dhanvantari, 800 B.C.

International intrigue: silent killers that are masters of disguise, blood hijackers, sex in exotic locations, and one of the richest men in the United States sending hordes of doctors to remote rain forests. The killers prey on pregnant women and young children. They have been around for more than 500,000 years. Only 5 years ago, they were responsible for more than a million deaths. And even though the forces of good are prevailing and those numbers are dropping rapidly, it will probably take more than 30 years to completely eliminate all of the blood hijackers. And what an ignominious end it might be, for it is quite possible that by genetically modifying the testicles of their unwitting steeds, the killers might be deprived of all the logistic support they need and thus be eradicated. If the glowing gonads don't do the job, there is an alternative plan to release millions of specially designed assassins armed with microinjections of scorpion venom that will surely finish off the invaders for good.

While much of this may sound like an excerpt from the latest spy thriller, it's not—it's all about malaria. The killers are parasites called *Plasmodium*, which ride around in mosquitoes, their weapon is malaria, and the rich man is Bill Gates.

More than 40 percent of the world's population is at risk of contracting malaria. (1) Almost 250 million people are infected each year, and it is estimated that in 2010 about 650,000 people died from the disease. According to the World Health Organization, "Malaria disproportionately affects poor people, with almost 60% of malaria cases occurring among the poorest 20% of the world's population." (2) Although the numbers are daunting, they are a vast improvement from a decade ago, when the numbers were higher and the outlook was dismal. A large influx of money, insecticide-treated bed nets, indoor spraying, and a new class of drugs called artemisinin-based combination therapies (ACTs) have reduced the death toll and number of infections faster than anyone expected.

Malaria is caused by protozoan parasites, *Plasmodium* spp., that spend their life oscillating between humans and mosquitoes. There are 120 species of *Plasmodium*, but only 5 of them naturally infect humans, *Plasmodium falciparum, Plasmodium malariae, Plasmodium ovale, Plasmodium vivax,* and *Plasmodium knowlesi*. The two most common malaria parasites are *P. falciparum* and *P. vivax*. The former is the more common and more deadly of the two and is found mostly in Africa, while *P. vivax* is more common in Asia. *P. vivax* is less lethal, but it hides in the human liver, where it is difficult to treat, and so it often escapes detection. *P. falciparum* can infect up to 80 percent of the victim's blood, 40 times more than *P. vivax*. Fortunately, the more virulent *P. falciparum* is much fussier than *P. vivax* and can live only in tropical mosquitoes.

P. malariae, P. ovale, P. vivax, and *P. knowlesi* cause cyclical fevers and anemia, which are debilitating but not fatal, whereas infection by *P. falciparum* is often lethal. The number of *P. falciparum* parasites in an infection is much higher than that for other malaria parasites, and, most important, only *P. falciparum* infections change the surface of red blood cells, which causes them to adhere to the side of blood vessels, disrupting organ function. These obstructions often occur in the brain and result in cerebral malaria.

From Mesopotamian tablets crafted around 6,000 years ago, to texts written by the ancient Hindu surgeon Dhanvantari in 800 B.C., to the writings of Homer, Plato, and Shakespeare, we can see that malaria has been responsible for pain and suffering as long as humankind has been recording its history. The ancient Chinese knew that malarial fevers were associated with an enlarged spleen and blamed malaria's headaches, chills, and fevers on three demons: one carrying a hammer, a second a pail of water, and the third a hot stove. (3) Having malaria is clearly a devastating experience.

P. falciparum was once commonly found in Italy, where the poet Dante and a number of popes were overcome by its fevers. The people of Rome were unaware of the connection between their sickness and mosquito bites, instead associating the disease with the unpleasant marsh gas smell bubbling up from wetlands, marshes, and stagnant waters. In their opinion, it was this bad air, *malaria*, that was responsible for the deadly disease. This was the origin of the word "malaria." The first time the word was used in the English language was in a 1746 letter by Horace Walpole, who wrote, "There is a horrid thing called malaria, that comes to Rome every summer, and kills one." (4, p. 69)

In 1880, Charles Laveran, a French army doctor working in a military hospital in Algeria, was the first person to observe malaria parasites in red blood cells of malaria patients. Two years later, he was posted to Rome, where he treated Italian soldiers also suffering from malaria. He examined the blood of 192 malaria patients and found the same parasites in 148 of the Italian samples as he had seen in the blood of the Algerian patients. Laveran was convinced the parasites were responsible for the malaria, particularly because he had also noticed that they disappeared from the blood when the patients were treated with quinine, an antimalarial drug, and went into remission. His reports that malaria was caused by a parasite were met with much skepticism, however, and it would take many years before the scientific world accepted the fact that the parasites discovered by Laveran were responsible for spreading malaria.

Fifteen years after Laveran discovered the malaria parasite, Amico Bignami, an Italian pathologist, was the first to publish the notion that it was the bite of a mosquito that transmitted malaria to humans. Not much later, Ronald Ross, a Scottish physician working in India, showed that mosquitoes transmitted Laveran's malaria parasite to birds. Giovanni Battista Grassi, a well-known evolutionary biologist, joined forces with Bignami and one-upped Ross by infecting a human volunteer with malaria parasites through the bite of an infected mosquito. They also showed that it was a very specific type of mosquito, the *Anopheles* mosquito, that carried the *Plasmodium*. Ross was much more vocal and eloquent than Bignami and Grassi, and he won a rather ugly war of words that led to his Nobel Prize in Medicine and Physiology in 1902 "for his work on malaria, by which he has shown how it enters the organism and thereby has laid the foundation for successful research on this disease and methods of combating it." In a short biography released by the Nobel Foundation after Ross received his award, the foundation found it necessary to write, "Whilst his vivacity and single-minded search for truth caused friction with some people, he enjoyed a vast circle of friends in Europe, Asia and America who respected him for his personality as well as for his genius." (5)

Although Laveran's research implicating the *Plasmodium* in causing malaria occurred 16 years before Ross's work, Laveran was not awarded the Nobel Prize for Medicine and Physiology until 1907, 5 years after Ross. Charles Louis Alphonse Laveran was awarded the prize "in recognition of his work on the role played by protozoa in causing diseases." According to the Nobel Foundation, "His first communications on the malaria parasites were received with much skepticism, but gradually, confirmative researches were published by scientists of every country"; in addition, "Laveran did not, for 27 years, cease to work on pathogenic Protozoa and the field he opened up by his discovery of the malarial parasites has been increasingly enlarged. Protozoal diseases constitute today one of the most interesting chapters in both medical and veterinary pathology." (6)

Malaria was once common in the United States and many European countries, but these areas have been free of the disease for a long time. Today, it is sub-Saharan Africa that is suffering more than any other region in the world. Not only is it the home of the most lethal malaria parasite of them all, *P. falciparum*, but also the parasite's host *Anopheles gambiae*, which does not hibernate, is the most aggressive of all mosquito species that are capable of harboring *Plasmodium* and the only mosquito to get its blood meals exclusively from humans.

There are 3,200 species of mosquitoes on earth, but the malaria parasite is transmitted by only 70 species, all of which belong to the *Anopheles* genus. The three main culprits are *A. gambiae* in Africa, *Anopheles stephensi* in Asia, and *Anopheles albimanus* in South America. Furthermore, because only the females go hunting and only the females have blood meals, they are the ones that spread malaria parasites among their human hosts.

The genomes of *A. gambiae*, *P. falciparum*, and *P. yoelii*, a malaria parasite that infects mice, were published in 2002. *P. falciparum* never makes life easy for humans, and deciphering its genome was no exception. It has 23 million nucleotide bases, with an unusually high proportion of adenine and thymine bases. This made cleaving the DNA and then reassembling the sequenced fragments very difficult computationally. *P. falciparum* has approximately 5,300 genes, and 60 percent of the genes code for proteins whose purpose and function we do not yet understand. Not only do we not know what many of the genes in *P. falciparum* code for, but there are no similar genes known in other species. (7)

Considering that so many people are dying from malaria and that about 5 percent of the world's population will be infected with malaria at some point in their lives, it is surprising that malaria research and control have been underfunded for most of the twentieth century. Fortunately, that changed around 2005 when some of the world's richest and most powerful citizens—Bill and Melinda Gates, U2's Bono, and Bill Clinton—lent their support to the antimalaria campaign, and antimalaria charities became a "hip way to show you care." The Gates Foundation alone spent more than $8 billion on global health issues between 1998 and 2008. Its campaign, which aims to not just control but eradicate the disease, has changed the way malaria is being attacked. As a consequence of this new emphasis, the annual expenditures to combat malaria increased from $100 million in 1998 to more than $1 billion in 2008. (1)

Today, just five Central African countries account for 50 percent of all global malarial deaths, and so it is not surprising that most of the efforts to control the disease are focused on sub-Saharan Africa, where the average person is exposed to hundreds of infective bites per year. The results have been positive. For example, in Zambia, malaria deaths have decreased by more than 60 percent since 2003. Efforts in Zanzibar and Rwanda have been just as successful. However, the larger, less peaceful countries in Africa are proving to be more difficult. Nigeria still accounts for more than a quarter of the African malaria cases, and malaria is the main cause of death in the Democratic Republic of the Congo. (8)

Dale Greenwalt works at the National Museum of Natural History in Washington, DC, but he regularly summers in Glacier National Park, Montana. He has shown that female mosquitoes have been dining on blood for more than 46 million years. Every summer Greenwalt and his wife collect roughly a thousand fossils, which they take back to Washington for cataloging and analysis. In one of their specimens they found a fossil of a female mosquito with an engorged stomach. According to Greenwalt, this was a very fortunate discovery. He explained, "The abdomen of a blood-engorged mosquito is like a balloon ready to burst. It is very fragile, and the chances that it wouldn't have disintegrated prior to fossilization were infinitesimally small." (9) Using mass spectroscopy and other analytical techniques, Greenwalt and his team were able to show that the female mosquito's abdomen contained significantly more iron and porphyrin, important building blocks of hemoglobin, than found in similarly fossilized male mosquitoes. The female mosquito had been snacking on someone's, species unknown, blood. (10)

As long as humankind has been recording its aches and pains, its dreams and desires, its scribes have been telling us about malarial fevers. But where did this disease originate? To find out where malaria came from, we have to go further back than the written word, but exactly how much further is not known. We are unlikely to find fossilized *Plasmodium*, so we have to resort to molecular biological techniques to answer that question.

Thanks to the sequencing of the *Plasmodium* genome, we assume that the malarial parasite was once an alga of some type that made its way into mosquito larvae. But when and how did the *Plasmodium* first infect humans? The origin of human infections by *P. falciparum* is the subject of interest of numerous research labs and the source of some controversy; many of our closest primate relatives have been accused of harboring the precursor of the human malaria parasite. Prior to 2010, it was assumed that *P. falciparum* has been perfecting the way it sneaks around the human immune system for the last 5 to 7 million years. At the time, the closest known relative to *P. falciparum* was the chimpanzee malaria parasite, *P. reichenowi*, and it was assumed that the two parasites diverged when the human and chimpanzee ancestors parted ways 5 to 7 million years ago.

Recently, however, other *Plasmodium* spp. similar to *P. falciparum* have been found in chimpanzees and gorillas. Beatrice Hahn, from the Department of Medicine at the University of Alabama, and her coworkers collected feces from 3,000 wild-living apes in Central Africa. Then some poor graduate students had the unfortunate job of sifting through all the feces looking for plasmodial DNA. No DNA from malaria parasites was found in the droppings of eastern gorillas or the bonobos, while approximately a third of all western gorillas and half of all chimpanzee feces that were sampled contained plasmodial DNA. The DNA sequences found in the western gorilla were the closest to human *P. falciparum*; we can therefore assume that *P. falciparum*, once upon a time, made the gorilla its home.

According to Hahn and her co-authors, "Among over 600 sequences derived from ape samples spanning most of Central Africa, we failed to find a single chimpanzee parasite that was sufficiently closely related to *P. falciparum* to represent a progenitor. Thus, *P. reichenowi*, as well as other chimpanzee *Plasmodium* species, can be excluded as precursors of *P. falciparum*. Instead, all new phylogenetic evidence points to a western gorilla origin of human *P. falciparum*." (11, p. 423) This means that *P. falciparum* and *P. reichenowi* did not coevolve with their hosts, which is a pity because now we have no idea when *P. falciparum* started infecting humans.

Studies such as these are very important because hopefully in the future they will allow us to predict when a pathogen will jump from its current host to humans. Furthermore, Hahn thinks her research "can inform malaria eradication efforts about potential zoonotic *Plasmodium* reservoirs and provide insights into adaptive changes that might be required for ape *Plasmodium* infection of humans." (*11*, p. 424)

No scientific theory is ever safe, and no sooner had Beatrice Hahn published her paper in *Nature* implicating the gorillas than Franck Prugnolle from the Institut de Recherche pour le Developpement in France and his coworkers pointed their fingers at the spot-nosed guenon. In a couple of papers in the *Proceedings of the National Academy of Sciences*, they described how they had taken blood samples from 338 monkeys in Gabon and found a spot nosed guenon infected with *P. falciparum*.(*12a*) The DNA of the malaria parasite differed from that of the parasite found in humans and was similar to what Hahn had found in gorillas. Prugnolle and his co-workers conclude, "Examination of the nuclear and mitochondrial genomes of this parasite reveals that it is specific of nonhuman primates, indicating that *P. falciparum*–related pathogens can naturally circulate in some monkey populations in Africa. . . . The capacity of *P. falciparum* to switch hosts represents a risk that new strains may be transferred from nonhuman primates to humans, and vice versa." (*12b*, p. 11948)

Hahn and her coworkers, however, are convinced that *P. falciparum* has its origins in the gorilla and have published a rebuttal. The debate about how and when *P. falciparum* started infecting humans is far from over, and many questions remain unanswered: not only do we not know which species harbored the malaria parasite before it made its home in humans, but we also don't know how *P. falciparum* made the cross-species jump.

A. gambiae is the vector for *P. falciparum*, and it gets its blood meals only from humans. This means that although *A. gambiae* is responsible for transmitting the malaria parasites among humans, we can't blame it for introducing the *Plasmodium* into *Homo sapiens*. That honor probably goes to our forefathers themselves, who liked to dine on other primates. Preparing and eating monkeys and gorillas was a messy, bloody business involving a significant exchange of bodily fluids, and it is easy to envision how a hungry hunter skinning his meal could end up with infected gorilla blood in his eyes, nose, or mouth. Like the HIV virus, *P. falciparum* probably entered humans through consumption and preparation of bushmeat. No matter when and where the first *Plasmodium* made the transition from primate to human host, there can be no arguing that the *Plasmodium* are highly evolved organisms, uniquely adapted to maximally exploit human beings for their reproduction.

Humans are complex hosts and are, of course, very different from mosquitoes, so the parasite has to make highly sophisticated adaptations in order to survive in both hosts. *P. falciparum* undergoes 10 morphological transitions in its life cycle, lives in five different tissues, propagates asexually in three of these tissues, and has to reproduce sexually at each transfer between hosts. The *Plasmodium* needs all 10 forms in order to evade the very different molecular defense mechanisms thrown at it by the mosquito and human immune systems.

When an infected female mosquito has a blood meal, between 20 and 100 *Plasmodium* infiltrate the human bloodstream, associating themselves with the salivary anticoagulation agents the mosquito injects into its victim. Figure 4.1 shows fluorescent *Plasmodium* in the salivary glands of a mosquito.

The mosquito needs to find hemoglobin in her victim's blood and feast on it so that she can nourish the eggs she is about to lay. The malaria parasites are also interested in the hemoglobin, which is located in the doughnut-shaped red blood cells, where it is responsible for blood's deep red color. Without hemoglobin, we would not survive—it carries oxygen from the lungs to cells throughout the body. Hemoglobin is so important to human survival that our bodies continuously break down old hemoglobin to make new hemoglobin. Something so vital to our well-being is bound to be well defended, and so it is. The human body does not make it

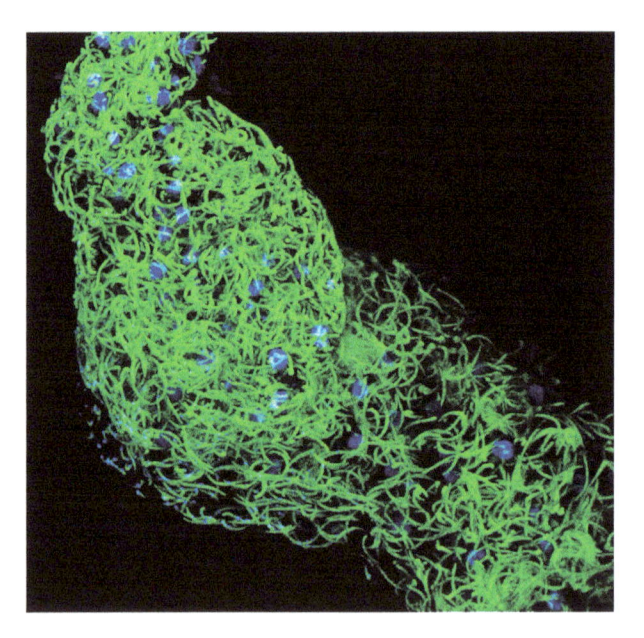

Figure 4.1 *Plasmodium* packed in a mosquito salivary gland. When the mosquito bites its victim, the green fluorescent malaria parasites move through the proboscis, enter the host's bloodstream, and begin the race to the liver. Even if the salivary glands of the mosquitoes contain more than 20,000 *Plasmodium* and they are massively drooling, no more than 1,000 parasites will be ejected at a time. (Image courtesy of Miguel Prudêncio.)

easy for the *Plasmodium* to gorge on our oxygen transporters. Before starting its bloodthirsty banquet, the malaria parasite has to disguise itself and amass an army. So, once in the human, the *Plasmodium* makes its way to the liver.

The trip from the mosquito bite to the liver is a hazardous one. From the moment they enter the host, the *Plasmodium* are chased by white blood cells. The pursuit can have unfortunate consequences for the infected victim. There is collateral damage to this microscopic chase because the malaria parasite can barge right through cells to avoid capture by the immune system's foot soldiers—the white blood cells. This almost always kills the cells. Fortunately, even though the parasites move 10 times faster than their pursuers and can barrel through surrounding cells, more often they get caught by the white blood cells because they move in a random pattern. (13)

The liver would not be on a "Top Ten Travel Destinations" list of organs in the human body; it is a waste treatment plant where all the unwanted molecules are sent for processing before being excreted. Security at garbage dumps is usually minimal or nonexistent, just as it is with the human liver. Once in its human host, the *Plasmodium* races to the liver to build up its forces and disguise itself before venturing back out. *P. vivax* takes extreme measures, making itself at home in the inhospitable human liver, where it can spend months and even years before entering the red blood cells. Because the first symptoms of a *Plasmodium* infection are observed only after the malaria parasites leave the liver, it is very difficult to diagnose malaria caused by *P. vivax*. The first symptoms in patients with *P. vivax* can appear months after exposure to the malaria-carrying mosquito. This can have some severe unforeseen consequences. For example, in a case reported in Germany, the liver of 20-year-old organ donor was transplanted into a 62-year-old woman who was suffering from cirrhosis. A month after the transplant, the women developed a high fever and exhibited malaria symptoms. Fortunately, her physicians were on the ball and were able to figure out that she had malaria. The patient had never been in a malaria endemic area, but the liver donor had been to Cameroon, where he was infected by *P. vivax* before dying from unrelated causes back in Germany. (14)

Fortunately, *P. vivax* is an exception, and most *Plasmodium* spend just over 5 days in the liver, since that is how long it takes to divide into about 50,000 new forms of the parasite, called *merozoites*. After being chased to the liver by the white blood cells, the merozoites are understandably reluctant to leave the relative safety of the liver. To improve their chances of survival in areas more carefully patrolled by the immune system, the parasites have evolved two impressive evasion techniques.

Robert Ménard, a researcher at the Pasteur Institute in Paris, his postdoctoral fellow, Rogerio Amino, and some co-workers have infected mice that have red fluorescent proteins in their arteries with green fluorescent *Plasmodium berghei* in order to establish how the malaria parasites protect themselves on the journey from the liver to the red blood cells. By photographing the green glow of the *Plasmodium* in the live mouse once every second, they were able to record the parasites' trickery. Before they leave the liver, the merozoites enter dying liver cells and disguise themselves in order to enter the bloodstream. They also prevent the liver cells from releasing a chemical signal that announces their upcoming death—a signal that would normally attract macrophages that would ingest the dead liver cells. (15, 16)

Once heavily disguised, the merozoites are then ready to start their quest for hemoglobin. The human immune system sees through this Trojan horse strategy, and most, but not all, merozoites are killed. The survivors invade the red blood cells, where they feast on the hemoglobin and asexually multiply until there is no room in the red blood cells. All the red blood cells burst synchronously, releasing waves of parasites that must find new hemoglobin-laden red blood cells to take over and pop. Unfortunately, the parasites are not well-behaved tenants. They excrete all their waste products into the red blood cells, causing them to rupture and spew the waste, which includes toxins like the hemozoin pigment, into the bloodstream. This is not good for the health of the human host, who can experience fevers, headaches, nausea, and total misery every time a new generation of parasites is released with the waste of its predecessors. The burst red blood cells can cause the spleen, which is responsible for clearing them out, to swell up to 20 times its normal size. Without treatment, the parasites increase by a factor of 10 every 2 days. The *Plasmodium* continues the destructive cycle until it is ready to leave the human host.

To prepare itself for the upcoming emigration from its human host, the *Plasmodium* changes form to a *gametocyte*, the sexual reproductive stage of the malaria parasite, in a red blood cell. Now it is ready to cruise the human bloodstream, circulating in the hope that an *Anopheles* mosquito will ingest the red blood cell it has commandeered. There may be close to a trillion *P. falciparum* in an infected individual, which ensures that at least some *Plasmodium* will enter a completely new life cycle within the mosquito.

While plasmodial reproduction in the human is asexual, the gametocytes that enter the mosquito after its blood meal are gendered and reproduce sexually. However, there is no romance for the *Plasmodium*—the best they can hope for is that the male and female gametocytes can hook up in the mosquito gut, for it is here that the sexual reproductive stage of the malaria parasite occurs. The male gametocyte penetrates the female, and a cyst is formed. The cyst protects the gametocyte while tens of thousands of sporozoites form inside it. This time, the malaria parasite does not try to trick its host's immune system; instead, it opts for brute force and just overwhelms it with sheer numbers. When mature, the cyst bursts, and multitudes of sporozoites swim up to the mosquito's salivary gland, where they wait for the mosquito to bite another human, and so the cycle starts again.

Mosquitoes are found in numerous climates, including the Arctic. A strain of *P. vivax* called *P. vivax hibernans* is found only in the Arctic. They have very short life spans during the brief northern summer. Rather than waste energy circulating in its human host, *P. vivax hibernans* hides out in the liver and starts circulating only after the mosquitoes in the swamps and puddles in the surrounding areas have hatched. It is speculated that some chemicals released by the first mosquito bite activate the *Plasmodium* in the liver and send them circulating through their host's blood, waiting for a mosquito to transport them to their next victim.

Maria Mota and Miguel Prudêncio, two well-known malaria researchers at the Institute of Molecular Medicine in Lisbon, Portugal, are particularly interested in understanding the interactions between the parasite and the molecular liver environment. They think that the *Plasmodium* parasites are vulnerable in the liver and that vaccines and drugs that target the parasite in the liver have a good chance of success. Additionally, if the infection can be eradicated at this point, the victims will never develop the symptoms of fever, anemia, cerebral malaria, and potential respiratory failure that accompany the blood stage of infection, and it will seem as if they have never gotten infected at all.

Mota and Prudêncio use hepatoma cell lines (cancerous liver cells) in their studies because they continually replicate and are good models for what happens in human liver cells. In order to find potential antimalarial drugs that target the liver and to investigate plasmodial behavior in liver cells, they used two strains of *Plasmodium* constructed and produced in the group of Chris Janse at the Leiden University Medical Center in the Netherlands, one of which expresses fluorescent proteins and the other firefly luciferase. This makes it possible to track parasites in the cancer liver cells, which can be cultivated and infected with the parasite, as well as in an infected mouse. Prudêncio says, "The luciferase-expressing *P. berghei* enables us to do in vivo studies of liver infection without sacrificing the mouse. It can be injected with luciferin and the bioluminescent signal of the infected liver can be quantified." (personal interview)

In one of their studies, conducted with the German biotech company Cenix, Prudêncio and collaborators placed the hepatic cells in separate wells, infected these cells with fluorescent *Plasmodium*, and systematically treated each well with one of 1,037 different FDA-approved drugs. The team found 13 different compounds that caused a decrease in the rate of infection. These compounds were then reassessed in vitro and in vivo using luciferase-expressing *P. berghei* parasites. One compound, decoquinate, was found to have an effect even at low doses and worked to inhibit the liver, blood, and the sexual stage of the parasite infection. (17) Mota and Prudêncio are continuing to investigate decoquinate as a potential antimalaria drug.

Malaria researchers have numerous tools to deal with their mosquitoes and keep track of the *Plasmodium*. In a recent tour of Miguel Prudêncio's lab, my students were particularly impressed and amused by two pieces of equipment that are commonly used by researchers in this field. The first was a cross between a handheld vacuum cleaner and a pipette, which the lab calls a bug vacuum; it is used to suck up individual mosquito larva. The second was a fluorescence-activated cell-sorting (FACS) cytometer, which can distinguish between fluorescent cells (that contain malaria parasites) and nonfluorescent, uninfected cells, count them, and even separate them from one another.

The malaria parasites are microscopically small, and it is difficult to track them even if they are fluorescent. There is no way to see where the *Plasmodium* is when the mosquito it is in is flitting about. The mouse and mosquito must be anaesthetized and kept stationary to find and follow the fluorescent parasites. The *Plasmodium* in the mice can be followed under a standard fluorescence microscope or a much more expensive 360-degree optical imager. *Scientist* magazine listed such a fluorescence detection system with 12 activating lasers and 48 detectors on rotating rings that can generate a full three-dimensional body scan of fluorescent proteins within a live mouse as one of its top 10 innovations for 2011. (18) Even with a high-end three-dimensional scanner, which can cost half a million dollars, it is very difficult to record the fluorescent individual *Plasmodium* unless they are very close to the surface of the skin; however, the parasites multiply rapidly, and the large numbers found in the liver are easily visualized.

Observing fluorescent *Plasmodium* as they infect the mouse during a mosquito blood meal is also challenging because the mosquito is unpredictable and it is impossible to know where and when it will bite the mouse. Catching the actual infection on film is much easier with slow feeders, such as ticks. In order to examine the interspecies transfer of fluorescent *Plasmodium*, researchers have developed some crafty tricks. To observe the parasites as they leave the mosquito, researchers superglue the infected mosquito to a glass slide. If this is done at a low enough temperature, the mosquito will be still and won't salivate. However, as soon as the temperature is raised, the mosquito will be energized and start salivating, thereby releasing fluorescent parasites in the saliva that can be counted. Freddy Frischknecht's research group from the University of Heidelberg Medical School, which is responsible for perfecting this technique, found that even if the salivary glands of the mosquitoes contain more than 20,000 *Plasmodium* and they are massively drooling, no more than 1,000 parasites will be ejected at a time. (*19*)

Plasmodium have also been created not to track the parasites but to see what is happening inside them. For example, green fluorescent protein has been linked to the actin in the actomyosin motor to allow the study of the mechanics of their movement within their mosquito hosts. (*20, 21*) For the first time, researchers have been able to see how the internal filaments of the gliding motor interact with the external environment of the parasite.

P. berghei is the most commonly studied *Plasmodium* species. The natural host of *P. berghei* is the African tree rat, and its vector is *Anopheles dureni*. It is particularly easy to work with because it does not infect humans and is a good model system for experiments that can't easily be conducted with *P. falciparum* and *P. vivax* in their human hosts. The fluorescence of individual GFP-expressing *P. berghei* can be detected in live mosquitoes. All three of the mosquito species responsible for transmission of human malaria can be infected with *P. berghei* by feeding them with blood from an infected mouse (figure 4.2). In all species of mosquitoes, the mouse malarial parasites float through the insect's blood and lymph systems,

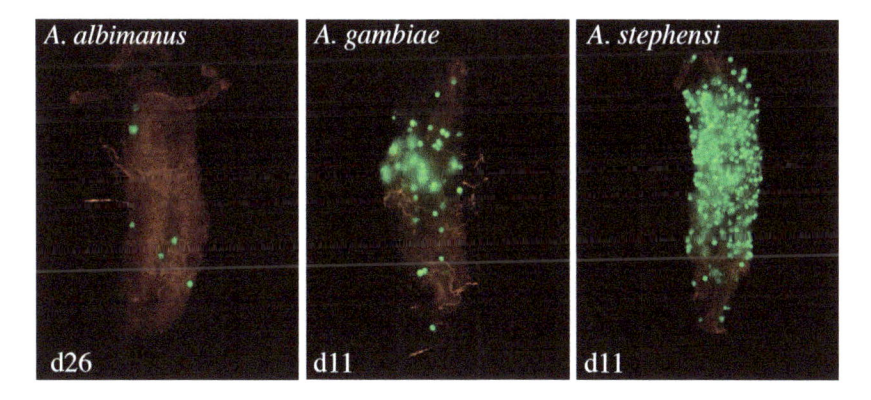

Figure 4.2 Photographs of the mosquitoes most commonly carrying the malaria parasite in South America (*A. albimanus*), Africa (*A. gambiae*), and Asia (*A. stephensi*) after they had a blood meal infected with fluorescent *P. berghei*. Individual *Plasmodium* can be distinguished (green dots) in these images of the mosquito midguts. (*22*)

invade its salivary glands, and are infectious in mice after subcutaneous injection. (22)

In 2003, Marcelo Jacobs-Lorena and his colleagues at the Bloomberg School of Public Health and Malaria Research Institute at the Johns Hopkins University showed that mosquitoes expressing SM1 proteins in their midgut do not transmit *P. berghei*. Before trying to find a protein like SM1 that would prevent transmission of the *P. falciparum* and *P. vivax* in *Anopheles* mosquitoes, the Hopkins researchers needed to establish whether the genetically modified mosquitoes would be able to compete with wild mosquitoes or whether the addition of the SM1 protein put them at a disadvantage. It took 3 years before they could report their results. In their experiments, 250 *A. stephensi* of one sex that had been genetically modified so that they expressed SM1 in their midguts and GFP in their eyes were released in a cage with 250 wild-type mosquitoes of the opposite sex. They were left to gorge themselves on *P. berghei*–infected mice and to reproduce. After 4 to 6 days, their eggs were collected, their larvae reared, and the pupae were placed in new mosquito cages, where they were fed with a 10 percent sugar solution. This procedure was repeated nine times, and for each generation the number of transgenic mosquitoes, those with fluorescent green eyes, and the wild-type mosquitoes were counted. And, yes, the transgenic malaria-resistant *A. stephensi* with the SM1 protein lived longer and outreproduced their brethren when fed on *Plasmodium*–infected blood. After nine generations, 70 percent of the mosquito population was resistant to *P. berghei*. (23)

Six years after Jacobs-Lorena and colleagues showed that transgenic mosquitoes can be created that do not carry the mouse malarial parasite *P. berghei*, Michael Riehle and his coworkers at the University of Arizona showed that *A. stephensi* can be modified so that it does not transmit *P. falciparum*. (24) It is only a matter of time before another laboratory finds the perfect combination of genetic modifications that will result in transgenic *Anopheles* that do not transmit any of the malaria parasites and can outreproduce naturally occurring mosquitoes. At that point, we will have to decide whether we want to release millions of genetically modified mosquitoes that are inhospitable to malaria parasites, so that they may eliminate malaria by replacing the current mosquito population, or whether we want to continue using environmentally harmful pesticides to control the mosquitoes.

An alternative strategy is to infect mosquitoes with naturally occurring bacteria that have been genetically modified to secrete antiplasmodium proteins whenever they are in the presence of the malaria parasite. An example of this technique was reported by Marcelo Jacobs-Lorena and colleagues in 2012 after their group had successfully infected *A. stephensi* with a strain of genetically modified bacteria that normally occur in the midgut of the mosquito. These bacteria were created to express GFP for monitoring purposes and small peptides, such as the aggressively named scorpine, that attacked the malaria parasites. The researchers found that the development of the human parasite *P. falciparum* and the rodent parasite *P. berghei* in the mosquito was suppressed by up to 98 percent. (25)

The malaria parasite has a sophisticated toolbox of genetic resources that allows it to continually reinvent itself to avoid our best efforts to control its infection. Consequently, we can't rely on the current methods of malaria control but must constantly adapt and think of new ways to curb the disease. To maximize our chances, we have to understand some of the evolutionary advantages malaria parasites and humans have garnered over time.

First, let's examine how the malaria parasite has evolved chemical tricks to manipulate the mosquito to do its bidding. While the parasites are multiplying in the mosquito's gut, they release a chemical that inhibits the desire of the female *Anopheles* to feed. Having fewer blood meals may not be good for her eggs, but the decreased exposure to an irate victim increases her chances of survival, and thereby the odds that the *Plasmodium* inside the mosquito will be squished by an annoyed human are reduced. Once the *Plasmodium* move into the mosquito's salivary gland, they are ready to invade the human bloodstream, they restore the mosquito's appetite, and simultaneously the wily parasites depress the amount of anticoagulants in the mosquito's saliva. As a consequence, during a blood meal the human victim's blood dries up before the mosquito gets enough, causing the insect to bite more often. Once again, this is not in her best interests, since each bite places her in jeopardy, but the *Plasmodium* get another chance to infect a new host. (4)

The mosquito parasites are not the only species to evolve; we the victims have also adapted to *Plasmodium* infections and malaria. More than 5 percent of the human population has genetic mutations that have evolved to fend off malarial infections. The most important of these mutations is responsible for the removal of proteins, known as *Duffy antigens*, from the surface of red blood cells. *P. vivax* is unable to infect hosts with this mutation. The mutation is so successful that for the last 5,000 years, all Central Africans have had it, and consequently *P. vivax* is no longer found in the rain forests of Africa.

The disappearance of *P. vivax* from Africa has left a void that was filled by *P. falciparum*. Given the severity of the infections spread by *P. falciparum*, it is not surprising that a new mutation has evolved in humans that protects them from *P. falciparum*. The mutation is found in the gene for hemoglobin, where it causes the red blood cells to collapse when attacked by the malaria parasite—they lose their doughnut shape and became rigid, sickle-shaped cells that are impenetrable by *P. falciparum* (figure 4.3). (26) The mutation doesn't come without its drawbacks, though; children of parents who both have the sickle cell mutation have a one in four chance of inheriting a copy of the mutated gene from each parent, which causes a fatal disease, sickle cell anemia. Normally a mutation that has such deadly consequences would not be passed on through many generations, but in the case of the sickle cell mutation, the protection against *P. falciparum* infections outweighs the consequences of a fraction of the population getting sickle cell anemia. To this day, the sickle cell mutation is found in up to 40 percent of the inhabitants of parts of Africa and South Asia. Fortunately, medical treatments are now available for individuals who have sickle cell anemia, and having two copies of the gene is no longer fatal.

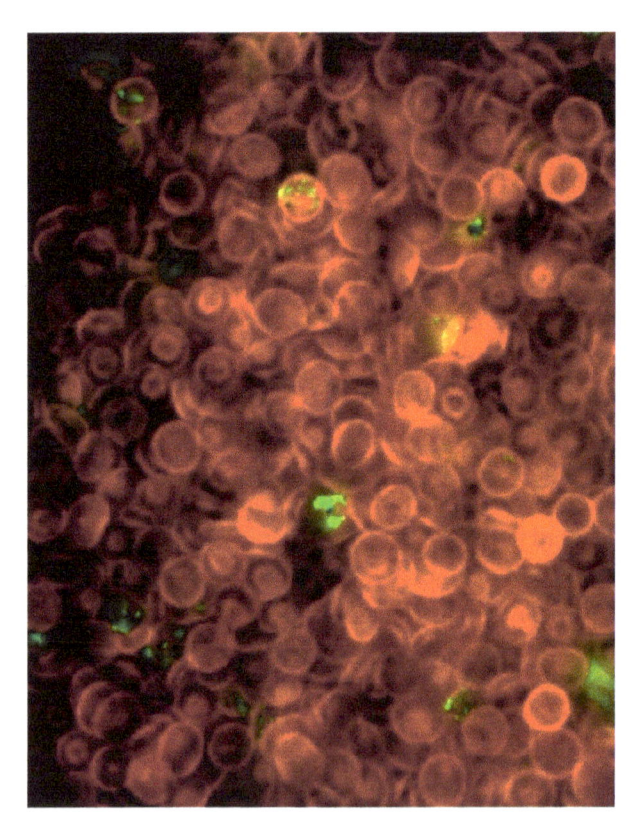

Figure 4.3 Some of the red blood cells shown in this photograph are infected with GFP-expressing *P. falciparum*. Because the patient has the sickle cell mutation, the cells infected by the malaria parasite collapse, protecting the human host against malaria. (26)

In the early 1950s, Anthony Allison was the first person to see the link between sickle cell anemia and malaria. At the time, it was a controversial finding because it was also the first time that someone had shown that diseases have a significant influence on human evolution. Nearly 70 years after Allison first reported that carriers of the sickle cell gene were protected against malaria, the molecular basis for the antiplasmodial activity of the sickle cell mutation was found. In humans, there is no secretory system for defective red blood cells. When they are infected or damaged, the cells are rapidly broken down by the spleen. However, the *Plasmodium* constructs a bridge that aids in clumping infected red blood cells together and sticking them to the walls of the blood vessels, thereby saving them from immediate dispatch to the spleen. Michael Lanzer and his colleagues at Heidelberg University in Germany and the Biomedical Research Center Pietro Annigoni in Ouagadougou, Burkina Faso, have used fluorescent proteins to show how the sickle cell mutation prevents the *Plasmodium* from building the bridge, thereby ensuring that the infected cells are dispatched to the spleen. Now that scientists know how the mutated hemoglobin prevents the inflammation associated with red blood cells sticking to each other and the cell walls of the blood vessels, new antimalarial drugs can be designed that mimic the processes observed in people with the sickle cell traits. (27)

Humans cannot rely on evolving immunity against the malaria-causing parasite, which reproduces 200 times faster than we do. It simply out-evolves us. To combat malaria, we have to go to the laboratory to design ways to kill all the *Plasmodium* parasites, kill all the *Plasmodium*-carrying mosquitoes, prevent the malaria parasites from being transmitted between humans and mosquitoes, or inoculate all humans so they cannot be infected by the malarial parasite. None of these are easy tasks.

Long before we knew that mosquitoes transmitted the malaria parasite, people have been using antimalarial drugs to combat the disease. Unfortunately, the *Plasmodium* parasites have often evolved resistance to the drugs faster than scientists could find or derive new drugs. Some drugs, such as quinine, have remained effective for hundreds of years, but that tends to be the exception. The earliest recorded applications of antimalarial drugs are mentioned in Chinese texts from 2700 B.C. that describe the use of plants to treat malarial fevers.

One of the biggest breakthroughs in the fight against malaria came in 1640, when it was reported that a tincture of the cinchona bark was commonly used to treat the disease in South America. By 1656, the British were drinking an infusion of the cinchona bark, but they were less likely to use the treatment than were the Spanish or Italians. In England, cinchona bark was known as "Jesuits' bark," and there was some resistance to drinking the hot, bitter remedy due to its strong association with Catholicism. Robert Talbor manipulated the situation to make his fortune. He was a self-educated doctor who saw the effectiveness of the unpopular cinchona bark brew in treating malaria fevers. Talbor concocted a mixture of cinchona, wine, and opium that was still effective but lacked the bitter taste of the original medicine. He sold this secret fever remedy as a safe alternative to Jesuits' bark. Word of his success at treating malarial fevers spread rapidly. Talbor was appointed the royal physician by Charles II in 1672, and he was knighted in 1678. He became the physician of choice to royalty all over Europe. Louis XIV of France paid Talbor 3,000 gold crowns and gave him a large pension and a title; in return, Talbor promised to reveal his secret formula upon his death.

The physician to the royal must have been a greedy man. He was not satisfied with his fame and fortune and clandestinely bought up all the cinchona bark to prevent any resurgence of the competition. Talbor's manipulations of the early pharmaceutical market did not last long, however; he died in 1681, at the age of 39. As promised, his secret was revealed to King Louis XIV. In 1682, an English translation document titled "The English Remedy: Or Talbor's Wonderful Secret for the Curing of Agues and Fevers—Sold by the Author, Sir Robert Talbor to the Most Christian King and since his Death ordered by His Majesty to be published in French, for the Benefit of his Subjects" was published, and Talbor's secret remedy was secret no more. For nearly a century it remained the drug of choice for the treatment of malaria until, in 1820, the active component, quinine, was extracted from the bark. Quinine has made *P. falciparum*'s life a misery for more than 400 years, but unfortunately malaria parasites have finally developed resistance to quinine. (28)

The newest generation of antimalaria drugs, the artemisinin-based combination therapies (ACTs), had their birth in the Vietnam War. During the war, the North Vietnamese soldiers lived, traveled, and fought in the mosquito-infested jungles of Northern Vietnam. They had no effective treatment against the malaria parasite, and many people died. On the other hand, only twelve American soldiers died from malaria. The *P. falciparum* found in the Vietnamese jungles was susceptible to quinine, and, although the drug had been known for more than a century and was in plentiful supply on the American side, it was not available to the North Vietnamese—hence the great difference in fatalities. Following the orders of Chairman Mao, Chinese scientists started an initiative to find a cure for malaria that could be used to help the North Vietnamese troops. Their search led them back to 168 B.C. and an ancient Chinese document called the "52 prescriptions," which described the medicinal properties of the sweet wormwood tree, *Artemesia annua*. The 2,000-year-old recipe became the basis of today's ACTs, which were used by the North Vietnamese to great effect in the last year of the war and subsequently reduced the number of malaria infections in China from 2 million to fewer than 100,000 a year. It took a while before ACTs were widely used outside of China because the Chinese kept their potent antimalarial drug hidden from Western researchers. It wasn't until 1979, when Chinese scientists published an English-language research paper about ACTs' efficacy, that Western researchers became aware of the drug.

Because most malaria victims cannot afford to pay for even the cheapest treatments, it took from 1979 until 1994 for a major Western pharmaceutical company to become interested in producing ACTs for the mass market. In 1994, Novartis acquired the rights for two critical components of ACTs, artemether and lumenfantrine, which have been used globally since 1999 and currently are the best antimalarial treatments available. ACTs have reduced the death toll and number of infections faster than anyone expected. However, as some malaria parasites build up resistance to ACTs, the drugs are losing their effectiveness, and it is time to find a new generation of antimalarial treatments.

P. falciparum parasites along the border of Thailand and Cambodia are showing signs of becoming resistant to artemisinin derivatives. (29) In today's world of international travel, it won't be long before a traveler infected with ACT-resistant malaria parasites flies to Africa or India and unleashes the ACT-resistant *P. falciparum*. This is a looming problem, and the Bill and Melinda Gates Foundation alone has spent $22 million to contain the spread of ACT-resistant malaria parasites from the Thai-Cambodian border. The malaria parasites from this region are notoriously potent; three of ACTs predecessors, chloroquine, sulfadoxinepyrimethamine, and mefloquine, all lost their potency in this region of Southeast Asia. The *Plasmodium* parasites here may be more genetically diverse, giving them an evolutionary advantage in their battle against antimalarial pharmaceuticals. Bed nets, especially those impregnated with artemisinin, are a cheap and effective method of limiting mosquito bites in Africa, but they don't work in Southeast Asia because the mosquitoes start biting earlier in the evening before people are home or in bed.

Artificial artemisinin derivatives were initially developed to provide an inexpensive alternative to artemisinin derivatives that are extracted from the sweet wormwood. Their molecular structures are dissimilar enough that ACT-resistant *Plasmodium* may still be susceptible to these artificial artemisinin derivatives.

Clinical and mathematical model studies have shown that the currently used drugs and insecticides will have to be supplemented with treatments that prevent the transmission of malaria parasites from infected patients to *Anopheles* mosquitoes if additional reductions in the incidence of malaria are required. David Fidock and his colleagues from the Division of Infectious Diseases at Columbia University have created *P. falciparum* that express firefly luciferase in such a way that the researchers can quantify the number of live parasites in each of the life stages inside the mosquito. They used their transgenic *P. falciparum* to screen a series of known antimalarial drugs, which help control the malarial outbreaks in humans, to find out whether some of them also inhibit the parasite in the mosquito, thereby preventing transmission between humans. Methylene blue had the highest transmission-blocking activity of any of the drugs tested, including the current first-line therapies, reducing parasite transmission to the *Anopheles* mosquitoes by 78 to 100 percent.

Methylene blue has an interesting history. It was synthesized and identified as an antimalarial drug by Paul Ehrlich in 1891 and was the first synthetic compound used in clinical therapy. Its use was discontinued because methylene blue turns urine green and turns the whites of the eyes blue. The compound's low price makes it attractive, and several drug trials are underway to combine methylene blue with other antimalarial drugs in the hope that they will negate methylene blue's colorful side effects. The fact that methylene blue also inhibits human-to-mosquito-to-human transmission by killing the *Plasmodium* just after they have entered the mosquito is sure to please the companies involved in these drug studies. (*30*)

Since 2007, when Bill and Melinda Gates called for a global effort to eradicate malaria, transmission-blocking vaccines (TBVs) have been the focus of a large scientific effort all around the world. The problem with TBVs is that the inoculated patient can't transmit malaria but can still get malaria. The vaccines don't provide a direct benefit to the recipient but instead offer protection to the community. Nevertheless, we can expect at least six TBVs to reach the clinical trial stage within the next 5 years. Clinical trials of any TBVs will be ethically challenging because vaccinated individuals aren't protected by the vaccines. According to Thomas Smith of the Swiss Tropical and Public Health Institute in Basel, a clinical trial would involve a number of communities with low levels of transmission. In half the communities, as many residents as possible will be inoculated with TBVs, while the remaining communities will be used as controls. According to Smith, "The readout would be whether transmission was interrupted in the vaccinated communities." (*31*, p. 848)

Traditional vaccines that are designed to protect the vaccinated individual have been under development for much longer than those designed to prevent malaria transmission. Individuals who have had repeated *P. falciparum* infections naturally gain an immunity to the parasite. The fact that the human body can build up resistance to the trillion blood-borne *Plasmodium* parasites has led to the development of vaccines designed to do the same. For more than 40 years, it has been known that immunity can be obtained by being bitten more than a thousand times by infected mosquitoes that have been weakened by irradiation. Although interesting, this observation had no practical application until 2013. After all, who wants to be bitten a thousand times? In 2013, highly positive results were reported for a vaccine that was designed to replicate the myriad of bites by weakened mosquitoes. The new vaccine, developed by Sanaria, a Rockville, Maryland, company, was 100 percent effective in very small trials. At this point only six people have been tested because the production of the vaccine is both technically demanding and laborious. The vaccine was created by growing mosquitoes in sterile conditions, infecting them by feeding them with *P. falciparum*–tainted blood meals, irradiating the mosquitoes to weaken the malaria parasites, and then harvesting billions of the weakened parasites from the mosquitoes' salivary glands. Initial research in which the subjects were subjected to dermal inoculations, similar to mosquito bites, was unsuccessful, but four intravenous doses of 135,000 weakened parasites did the trick. Mass vaccinations by intravenous injections are unlikely even if the production of the vaccine can be streamlined and its storage, requiring liquid nitrogen, simplified. According to Adrian Hill, a malaria researcher at the Jenner Institute in Oxford, England, "It's very unlikely to be deployable in infants or young children." (32) Stephen Hoffman, the lead author of the Sanaria study, says this is not a "show-stopper" because the 0.5 milliliters required to inject 135,000 parasites is a tiny volume, which will allow Sanaria to examine other ways to improve vaccine delivery.

The Sanaria vaccine is unusual in that it uses whole parasites; most other vaccines use certain plasmodial proteins. Hoffman and his co-authors explain, "We speculate that the vaccine induces immunity to a broad spectrum of antigens among the ~1,000 expressed by attenuated malaria parasites. The capacity to induce a cascade of multiple adaptive and innate effector cells and the breadth of antigenic specificity may explain the advantage of whole parasite vaccines compared to existing subunit vaccines." (33, p. 1364)

While most vaccines and antimalarial drugs target the *Plasmodium*, there are other equally valid methods to control or eradicate malaria that focus on the mosquito. Although there are over 3,000 different species of mosquitoes on earth, the malaria parasite is transmitted by only 70 species, all of which belong to the *Anopheles* genus. The most aggressive of these is *A. gambiae*, which is unfortunate because they carry *P. falciparum*. An obvious way of controlling the spread of malaria is to control the number of malaria-carrying mosquitoes or, if that is not possible, then the number of mosquito bites. However, this is not trivial; feeding on human blood is critical to the female *Anopheles*, which needs the blood to nourish its eggs and will do anything to get it.

The most effective method of mosquito control to date has also been the most controversial one. In 1939, Paul Müller, a chemist at the pharmaceutical company Geigy in Basel, Switzerland, discovered that DDT was an amazing insecticide, as very small amounts were required to kill insects, one application would last for months, and it did not seem to affect humans. When asked about the insecticide, Winston Churchill said its effects were miraculous, and he was right. DDT affects nerve signaling in all insects and has been used to control body lice that transmit typhus, plague-carrying fleas, and mosquitoes that carry malaria and yellow fever. Thanks to the widespread use of DDT in World War II, it was possible to prevent a repeat of the 5 million typhoid deaths that occurred in World War I. Consequently, Müller received the Nobel Prize in Medicine and Physiology in 1948 for his "for his discovery of the high efficiency of DDT as a contact poison against several arthropods."

During the Second World War, American forces in the South Pacific dipped their mosquito nets into a 5 percent solution of DDT, and so insecticide-impregnated nets were developed. Even though the insecticides used have changed over the years, impregnated nets are still one of the most effective techniques for preventing the spread of malaria.

From the 1950s to the 1970s, DDT was aggressively used to eliminate malaria mosquitoes by spraying it indoors and over mosquito breeding sites. When DDT is sprayed against building walls, it kills the mosquitoes that land on the sprayed wall for up to 6 months after application. It is estimated that the increased use of antimalarial drugs and DDT spraying prevented at least 500 million people from contracting malaria. However, DDT's days were numbered, as it was widely overused during the 1950s and 1960s, especially in agriculture. In the United States, between 70 and 80 percent of the DDT produced was used in farming. More than 600 million kilograms of DDT were used in the United States alone, and more than 1 billion worldwide. Although most of the DDT was used in agriculture, particularly in cotton fields, mosquitoes were exposed to it, and in many areas they built up immunity. Furthermore, most of the agriculturally used insecticide was washed off the fields into lakes and rivers, where it entered the food chain, primarily affecting large birds, such as the osprey, that lived off the fish they caught. The DDT made their egg shells so brittle that they broke easily.

In 1962, Rachel Carson published the book *Silent Spring*, which traces how DDT travels up the food chain. Its title refers to the possibility of a spring devoid of birdsong, particularly the American robin, whose numbers were decreasing due to its consumption of DDT-laden earthworms. The book has been credited with the 1972 ban of DDT in the United States and with starting the modern environmental movement. The United Nations has classified DDT as a persistent organic pollutant; it has been banned in the United States, Europe, and Canada, and its use is severely restricted in Mexico, India, and China. Its use is not completely prohibited because DDT is still one of the most effective ways of controlling malaria mosquitoes, especially when used correctly. Although DDT is not banned in sub-Saharan Africa, where malaria is endemic, its use has been severely curtailed because nongovernmental organizations and wealthy donor nations are reluctant to fund projects involving the pesticide. This is slowly changing, however. For example, in 2007 the World Health Organization reversed its policy of not funding projects involving indoor antimalaria DDT spraying. (*34*)

The ban of DDT may be hampering the campaign against malaria, but it has had the desired effect in the United States, where the bald eagle and osprey, once on the edge of extinction, are commonly seen again.

The use of DDT was so widespread, and it is so persistent, that we all have measureable amounts of it in our bodies. Fortunately, it seems to have no health consequences in humans (although there have been some claims that it causes cancer), and the amounts are decreasing due to the ban in most countries. It is not hard to see why DDT has stirred up so much controversy. In deciding whether DDT should be used, regulators have weighed its utility in combating malaria mosquitoes against its effect on birds and its discovery in humans all over the world and have decided to selectively ban and restrict its use to areas where malaria is a significant problem.

To date, most of the mosquito control methods have focused on indoor mosquitoes. The combined use of indoor spraying with DDT and pesticide-impregnated bed nets has been shown to be very effective. However, the mosquitoes are just as wily as the *Plasmodium*. There are two forms of *A. gambiae*, an indoor and an outdoor resting form. (*35*) In order to control malaria outbreaks and to have any chance at eradicating malaria, both forms will have to be considered. Eliminating the indoor mosquitoes will just favor the outdoor mosquitoes, which are much more abundant than previously thought and are highly susceptible to infection by the human malaria parasite.

While trying to eradicate mosquitoes that can carry *Plasmodium* is certainly desirable, one has to admire the female mosquito, if only for the risks she takes in order to get her blood meal. Even as children, we recognize the irritating buzzing sound a mosquito makes when it flies around. We associate the noise with an upcoming bite and are ready to kill the irksome insect the moment it lands somewhere. So, to avoid detection, mosquitoes feed at dusk or at night. While somehow avoiding a swatting hand, the mosquito has to pierce our skin. This would be a painful pinch, which would lead to her certain detection if the mosquito didn't numb the area with an injection of a painkiller excreted by her salivary gland. The salivary injection also contains an anticoagulation agent and is the vehicle of choice for the malaria parasite's entry into the human bloodstream. The feeding mosquito sucks up the thick and viscous human blood until she has ingested several times her body weight and can barely fly away. Ready to pop, she gets her bloated body to the nearest vertical surface. There she has to sit for 45 vulnerable minutes while she excretes all the water in the blood meal she sucked up. The whole process is so dangerous that the male mosquito never has a blood meal, and the female only indulges herself to nurture the eggs she is carrying. Males and nonpregnant females satisfy their hunger with plant nectar. (4)

How often have you heard someone say, "Mosquitoes just love me to death" or "If there is a mosquito anywhere within 10 miles of me, it will find me and bite me"? Anthony Brown was one of the first people to try to find out whether there was any truth to these statements. In the 1950s, he built some metallic scarecrows, dressed them up, and took them into the woods, where they were meant to fool the mosquitoes into thinking that they might be the source of a good blood meal dinner. Sounds crazy, but it worked. Brown, who at the time was at the University of Western Ontario in Canada, found out that many more mosquitoes landed on his humanoid mosquito lures when they were warmed up to body temperature and when they exhaled carbon dioxide. Cold scarecrows were not appealing to the Canadian mosquitoes. Brown also found out that wearing a wet sweater was not good for his iron men; not only would they rust, but the mosquitoes in the surrounding woods were attracted by the wet garments, particularly if they were soaked in human sweat.

Brown's findings have stood the test of time. Most mosquitoes are indeed attracted to puffs of carbon dioxide, moisture, and heat. However, that does not explain why some mosquitoes, such as *Aedes aegypti*, which transmit dengue fever, and *A. gambiae* feast almost exclusively on humans. How do they distinguish between Brown's metal scarecrows, a nice juicy cow, and a human? That question was answered nearly 50 years after Brown's work, when Bart Knols, a researcher at the Waginingen Agricultural University in Holland, observed that *A. gambiae* preferred feeding on its victim's lower legs and feet, even when the whole body was exposed. This contrasted dramatically with the Dutch mosquito species *Anopheles atroparvus* that prefers biting in the facial region. Theorizing that *A. gambiae* might be attracted by the stinky foot smell, Knols found a substitute—Limburger cheese. Waft some Limburger cheese odors near *A. gambiae*, and they will come flying, while other mosquito species that don't exclusively bite humans are not so interested in the smelly foot odors. Knols's choice of Limburger cheese was a good one because it is made using *Brevibacterium linens*, a close relation of *Brevibacterium epidermis*, the bacteria that make their smelly homes in the warm and humid spaces between our toes. (36)

Mosquitoes don't have noses; instead, their antennae are their primary olfactory organs. Numerous research groups all over the world are interested in understanding the intricacies of the mosquito's ability to smell. Many are using genetically modified mosquitoes to answer their questions. There are mosquitoes with fluorescent neurons leading to the olfactory organs in their antennae, as well as mosquitoes that have the proteins that are responsible for binding the odorant molecules tagged with fluorescent proteins. Different smelly molecules, such as those released by stinky feet or sweaty armpits, bind different odor receptors; as they bind, more receptor proteins are expressed, and since these are tagged with fluorescent proteins, researchers can observe the "where and when" of mosquito smelling on a molecular scale.

Female mosquitoes find their human subjects by following the smelly foot odor, as well as the carbon dioxide and lactic acid emitted from skin, sweat, and breath. It would seem that the malarial parasite has one more trick up its sleeve—it induces people infected with malaria to give off an additional odorant. This odorant makes biting infected humans more attractive to mosquitoes than biting healthy humans and improves the chances of transmission to other humans.

The commonly used mosquito repellant diethyltoluamide (DEET) targets the mosquito's olfactory system. Unfortunately, DEET is expensive, and resistance to DEET has been observed in some mosquito species; an alternative is needed. Stephanie Turner and her coworkers at the University of California, Riverside, have discovered new molecules that can disrupt the ability of the female mosquito to follow the puffs of carbon dioxide we exhale. These inhibitory molecules were isolated from ripe fruits. (37)

Normally, fruit flies release carbon dioxide to alert other fruit flies of danger. Earlier work by Turner, however, had shown that fruit flies are attracted to ripe fruits, despite the fact that the fruits emit large amounts of carbon dioxide. (38) How do fruit flies distinguish between carbon dioxide released as a warning and carbon dioxide associated with ripening fruits? Turner found that ripe fruits also release inhibitory odorants that prevent the fruit flies from detecting the carbon dioxide they release. This finding could be very useful, according to Turner, since "the carbon dioxide-response-modifying odors offer powerful instruments for developing new generations of insect repellents and lures, which even in small quantities can interfere with the ability of mosquitoes to seek humans." (39, p. 40) In 2013, Olfactor laboratories started an online campaign to attract investors to fund the testing of "Kite," a patch manufactured using the fruit fly inhibitor discovered by Turner. The patch has a shelf life of 18 months and works for 48 hours.

In the last 5 years, transgenic techniques have reached the point where they have the potential for being used in the fight against malaria. Mosquitoes have been genetically engineered to reduce plasmodial multiplication in their guts, to prevent the malaria parasites from living in the mosquito, and to collapse the *Anopheles* mosquito populations. In all cases, the genetically modified mosquito population has to replace the wild mosquitoes that carry the malaria parasite. This is not trivial, for if these genetically modified mosquitoes were ever released, the billions of wild mosquitoes that buzz around India and equatorial Africa would massively outnumber them. To have any effect on the spread of malaria, they will have to outlive and outbreed the wild-type bloodsuckers, and the genetic changes responsible for their malaria resistance have to spread rapidly through the mosquito population.

Selfish jumping genes spread rapidly by forming copies of themselves throughout the genome. Researchers hope to give the genes designed to combat malarial infections an advantage by linking them to selfish genes so that they have the edge they need to overcome the breeding advantage the wild-type mosquitoes have over transgenic males. Alternatively, the selfish genes can act as molecular guerrilla fighters that have been designed to enter the mosquito's cells, where they sabotage genes key to the malaria parasite's well-being in the mosquito.

Andrea Crisanti of Imperial College in London has used selfish genes in a proof of concept experiment that demonstrates how the gene necessary for housing *Plasmodium* in the mosquito can quickly and thoroughly be disrupted in a collection of suitably genetically modified mosquitoes. In an article published in *Nature* in April 2011, Crisanti and coworkers describe how they created a selfish gene that targeted and destroyed the GFP gene in the mosquito's sperm. Because the selfish gene sneaks into the genome of all the mosquito offspring, it spreads very quickly. To test their ideas, the researchers created a population of GFP-expressing mosquitoes (figure 4.4). The fluorescent glow of the green critters was then gradually turned off by the release of a small number of mosquitoes with the GFP targeting selfish genes. After 12 generations, more than 60 percent of the GFP genes were compromised. When larger populations with the selfish gene were released, the GFP fluorescence was extinguished even faster. (*40*) According to Crisanti, "The next step is to make the selfish gene break up not the fluorescent protein gene but one that is crucial for malaria transmission. It could be an odor-recognition gene that helps the mosquito finds its host, for instance, or one that the malaria parasite needs to enter the mosquito's salivary glands; the team already has 10 to 15 candidates." (*41*)

Another potential method for preventing malarial transmission is the use of a specially designed plasmodial assassin that kills the malaria parasite by injecting scorpion venom into infected mosquitoes. The finely crafted killer, *Metarhizium anisopliae*, is a parasitic fungus. When its spores land on a mosquito, they drill through the insect's cuticle to invade it. Raymond St. Leger, a biochemist and pathologist at the University of Maryland, College Park, and coworkers have genetically modified the

Figure 4.4 Mosquito larvae expressing GFP will turn into green fluorescent mosquitoes. When these mosquitoes breed with other mosquitoes that have a selfish gene, which targets the GFP, their offspring are most likely non-fluorescent. (*40*) (Image credit Sinclair Stammers.)

fungus with genes for an antimicrobial protein originally found in scorpion venom, a human antimalaria antibody, and a protein that blocks the *Plasmodium*'s access to the mosquito's salivary gland. In one final genetic modification, the parasitic fungus was transformed into a *Plasmodium* killer extraordinaire. A promoter was added to *Metarhizium anisopliae* so that it expresses its three new proteins only after the fungus had drilled into the mosquito's hemolymph, where the *Plasmodium* circulates. In 2011, St. Leger and colleagues reported on experiments in which *A. gambiae* heavily infected with *P. falciparum* were sprayed with the genetically modified fungus, and 98 percent of the *Plasmodium* parasites were eliminated by the fungus injections. (42)

Another method of controlling mosquito populations is based on the fact that female mosquitoes have sex only once in their lifetime. If the one and only time they mate with another mosquito happened to be with a sterile male, then that female would produce no offspring. This idea of controlling pest populations in which the female has sex a very limited number of times—an approach called the *sterile insect technique*—has been in use since the mid-1950s. Every year hundreds of millions of sterilized male screwworms and Mediterranean fruit flies are released. The males can have intercourse more than once and will spend their lives looking for available females. If enough sterile males are repeatedly released, they can collapse the local population of the pest.

In 1959, in Florida, the screwworm was the first species to be successfully controlled by sterile insect techniques. It has been completely eliminated from the United States, Mexico, and most of Central America. The primary screwworm, or blowfly, has a four-stage life cycle: egg, larva, pupa, and adult. The female adult is predacious and will feed on fluids from open wounds. She lays between 100 and 400 eggs in open wounds or in oral, anal, or nasal areas of its hosts—locations that will provide the larvae with food, warmth, and moisture. Upon hatching, the larvae burrow their way into the host, eating the surrounding flesh. They can break through the skin but generally don't need to do so. The primary screwworm larvae, *Cochliomyia hominivorax*, typically infest and eat the flesh of warm-blooded animals such as cows and occasionally humans. The first signs of an infestation are a reddish brown discharge that leaks from the wound, occasionally accompanied by an unpleasant odor. After 5 to 7 days of feasting on the host's flesh, the maggots eject themselves and drop to the ground, where they form the pupal life stage in the topsoil. Although they are largely under control in the Americas, screwworms are still major pests in Africa. The largest known screwworm outbreak occurred in 1989, affecting more than 2.7 million sheep in northern Africa. A screwworm "factory" in

Mexico produced 1,260 million sterilized male blowflies that were sent to the site of the infestation and released to successfully control the outbreak.

Because the larvae must breathe, they don't burrow very deeply into the flesh. Nevertheless, treatment of a blowfly infestation is an unpleasant process. The maggots need to be removed with tweezers, and the dead flesh must be removed. This is a painful procedure that requires antibiotic treatment. While the primary screwworm prefers live, warm-blooded hosts, the secondary screwworm (*Cochliomyia macellaria*) larvae prefer dead bodies and carcasses, so they are often used in forensic investigations. That's a bit too much screwworm information in a chapter about malaria, but it is hard to leave them out, seeing as they are so nasty and provide an excellent introduction to sterile insect techniques.

Sterile insect techniques are an excellent way of controlling a pest population because they eliminate the use of environmentally harmful pesticides and are species specific. For example, the Mediterranean blowfly will mate only with other Mediterranean blowflies. The main requirements for the sterile insect technique to work is that the female insect mates only once, while the male can mate numerous times, and the sterilized male has to be mobile so that he can find numerous partners.

The *Anopheles* mosquitoes meet these criteria: during copulation, certain proteins and peptides released from male accessory glands inhibit future mating behavior in female mosquitoes, yet field experiments with sterilized males have only recently begun. There are three reasons that attempts to eliminate mosquitoes by sterilizing the males are lagging more than 50 years behind the very successful screwworm elimination program. These are the need to find a way to determine the sex of the mosquito larvae; a way to overcome the reduced male evolutionary competitiveness due to the sterilization procedure; and the desire to have a method for monitoring the survival and dispersal of sterile males in the field. It is a trivial process to distinguish between the adult female and male mosquitoes. However, it is completely impractical to catch a million mosquitoes, sex them, separate them, kill the females, and sterilize the males. To be practical, one needs to find a way to sex the mosquito larvae because they are immobile and determining the sex and separating the sexes can be mechanized. This is what is done with the screwworm and Mediterranean fruit fly larvae. Unfortunately, until recently, there was no way to distinguish between male and female *Anopheles* larvae. In the case of the screwworm, the sterilization and accidental release of an occasional female has little effect on the efficiency of the program, but the release of sterilized female mosquitoes would be worse because they could still bite humans and transmit the malaria parasites. In order to successfully control mosquito populations by sterile insect techniques, a method to reliably sex mosquito larvae needed to be found.

In 2005, Andrea Crisanti, whose lab was also responsible for the selfish gene experiments, described the creation of genetically modified *A. stephensi* that could be used in sterile insect technique releases. *A. stephensi* are the most commonly found malarial mosquito in Asia, and they are easier to genetically manipulate than *A. gambiae*. The mosquitoes were genetically modified so that all mosquitoes expressed DsRed, a red fluorescent protein, and the male larvae expressed GFP in their testes (figures 4.5 and 4.6).

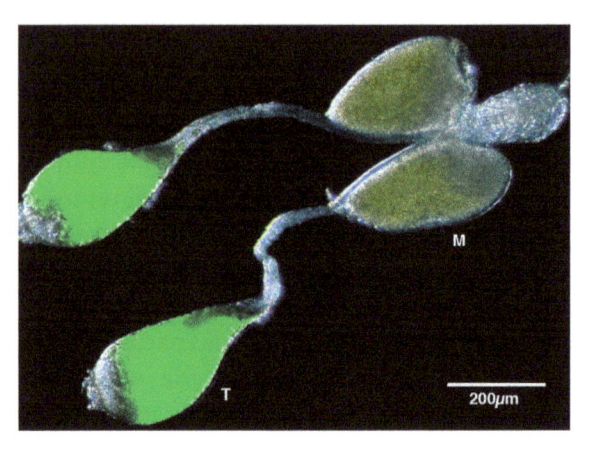

Figure 4.5 The internal reproductive organs of a male *A. gambiae* transgenic mosquito. The male accessory glands (M), in which seminal secretions containing proteins and peptides that inhibit future mating behavior in females are produced, and the testes (T), where sperm cells develop, are indicated. (*43*)

Figure 4.6 Two male mosquito larvae. Their sex is easily determined by noting the presence of a pair of green fluorescent testes. (*44*)

These genetic modifications made it possible to distinguish genetically modified mosquitoes from wild-type mosquitoes as well as females from males. The green fluorescence is detectable in the larval stage and is 100 percent accurate. Females are not influenced by the fluorescent testes in choosing a partner, which is not surprising because they would need blue light and a blue light filter to see the green fluorescent gonads, and genetically modified males are not impaired in their mating ability. Using a high-throughput sorter, the authors were able to separate more than 18,000 genetically modified males and females per hour. The glowing gonads have additional advantages—they allow researchers in the field to track the released mosquitoes, and they produce fluorescent sperm that can be tracked in the female spermatheca, the female sperm storage organ. By tracking the fluorescent sperm in caged females, the researchers were able to show that males could copulate with two to three females each, which is consistent with mating behavior in wild-type mosquitoes. Two of the major hurdles to using sterile insect techniques on malaria mosquitoes have been overcome. Mosquito larvae can be efficiently and accurately sexed, and the sterilized males can be distinguished from the fertile wild-type males in the field. The final hurdles that need to be overcome are the creation of competitive sterilized males and an acceptance by the public and governments to release hundreds of millions of genetically modified mosquitoes with glowing green gonads. (43, 44)

Sterilized insect techniques are species-specific pesticides, which makes them very effective and environmentally friendly methods. Every month more than a billion sterilized insects are released to control a large variety of pests. Before they are released, the sterilized insects have to be marked.

As its name implies, the cotton pest moth destroys cotton plants and is not very popular with cotton farmers. For decades, sterile insect techniques have been used to control the moths over millions of acres of cotton plantations. Feeding their larvae with a dye marks sterilized moths. Occasionally, the dye fades and is hard to see. This can lead to expensive reactionary overreleases of sterilized moths. Taking a page from their dengue mosquito research, scientists at Oxitec have created sterilized male cotton pest moths that express red fluorescent proteins and have shown it to be a much more reliable marker than the currently used dyes (figure 4.7). (45)

The war on malaria has intensified. The Gates Foundation is leading the charge by funding the distribution of impregnated mosquito nets, and new and old antimalarial drugs, as well as the development of new methods to eradicate malaria. It is unlikely that one battle with the

Figure 4.7 Two cotton pest moths shown under normal light on the left and under blue light on the right. The moth on the right has been genetically modified to express the red fluorescent coral protein DsRed2. The red fluorescence can be used as a marker to identify sterilized males released in cotton plantations during cotton moth control drives by sterile insect techniques. (45)

malaria parasites will suffice; it will have to be a carefully planned campaign involving attacks from numerous fronts. Old and tried, but no longer very effective methods, such as DDT and quinine, will have to be used in conjunction with seemingly off-the-wall ideas, like mosquitoes with glowing gonads and fungi that inject scorpion venom into mosquitoes. The increase in funding for malarial research since 2000 has had a very positive effect on the spread and control of the disease, but there is some debate on the next steps required to eradicate malaria. One group of scientists and public health officials would like to eradicate malaria in countries like Rwanda, Namibia, and South Africa, thereby shrinking the areas where malaria is found. Others disagree and are worried that this strategy will divert desperately needed funds from the countries that need the most help, like Nigeria and the Democratic Republic of the Congo.

To date, our track record at eradicating disease has not been very good. The malaria parasites are wily opponents that quickly adapt to any obstacles we place in their way. It will require significant global effort to continue beating back malaria and to prevent it from taking advantage of the increased spread of its hosts, the *Anopheles* mosquitoes, whose range is increasing due to global warming. But if we could add malaria to polio and smallpox, two diseases that have been conquered, it will have been a war worth fighting.

REFERENCES

1. Kappe, S. H. I., Vaughan, A. M., Boddey, J. A., and Cowman, A. F. (2010). That was then but this is now: Malaria research in the time of an eradication agenda. *Science 328*, 862–866.
2. World Health Organization. (2006). WHO gives indoor use of DDT a clean bill of health for controlling malaria. www.who.int/mediacentre/news/releases/2006/pr50/en/.
3. Babu, B. V. (2007). History of malaria and antimalerials. *RMRC News Bulletin*, 1–4. Bhubaneswar: Regional Medical Research Centre.
4. Shah, S. (2010). *The fever: How malaria has ruled humankind for 500,000 years.* New York: Farrar, Straus and Giroux.
5. Nobelstiftlesen. (1964). *Physiology or medicine.* Amsterdam: Published for the Nobel Foundation by Elsevier and http://www.nobelprize.org/nobel_prizes/medicine/laureates/1902/ross-bio.html.
6. Alphonse Laveran—Biography. http://www.nobelprize.org/nobel_prizes/medicine/laureates/1907/laveran-bio.html.
7. Plasmodium genomics. (2002). Special issue, *Nature 419*.
8. Enserink, M. (2010). Redrawing Africa's malaria map. *Science 328*, 842.
9. Yong, E. (2013). Blood-filled mosquito is a fossil first. *Nature News* doi:10.1038/nature.2013.13946.
10. Greenwalt, D. E., Goreva, Y. S., Siljestroem, S. M., Rose, T., and Harbach, R. E. (2013). Hemoglobin-derived porphyrins preserved in a Middle Eocene blood-engorged mosquito. *Proceedings of the National Academy of Sciences of the United States of America 110*, 18496–18500.
11. Liu, W. M., Li, Y. Y., Learn, G. H., Rudicell, R. S., Robertson, J. D., Keele, B. F., Ndjango, J. B. N., Sanz, C. M., Morgan, D. B., Locatelli, S., Gonder, M. K., Kranzusch, P. J., Walsh, P. D., Delaporte, E., Mpoudi-Ngole, E., Georgiev, A. V., Muller, M. N., Shaw, G. M., Peeters, M., Sharp, P. M., Rayner, J. C., and Hahn, B. H. (2010). Origin of the human malaria parasite Plasmodium falciparum in gorillas. *Nature 467*, 420–425.
12a. Prugnolle, F., Durand, P., Neel, C., Ollomo, B., Ayala, F. J., Arnathau, C., Etienne, L., Mpoudi-Ngole, E., Nkoghe, D., Leroy, E., Delaporte, E., Peeters, M., and Renaud, F. (2010). African great apes are natural hosts of multiple related malaria species, including Plasmodium falciparum. *Proceedings of the National Academy of Sciences of the United States of America 107*, 1458–1463.
12b. Prugnolle, F., Ollomo, B., Durand, P., Yalcindag, E., Arnathau, C., Elguero, E., Berry, A., Pourrut, X., Gonzalez, J.-P., Nkoghe, D., Akiana, J.; Verrier, D.; Leroy, E., Ayala, F. J., and Renaud, F. (2011). African monkeys are infected by *Plasmodium falciparum* nonhuman primate-specific strains *Proceedings of the National Academy of Sciences of the United States of America 108*, 11948–11953.
13. Amino, R., Giovannini, D., Thiberge, S., Gueirard, P., Boisson, B., Dubremetz, J. F., Prevost, M. C., Ishino, T., Yuda, M., and Ménard, R. (2008). Host cell traversal is important for progression of the malaria parasite through the dermis to the liver. *Cell Host and Microbe 3*, 88–96.
14. Wolfe, N. (2011) *The viral storm: The dawn of a new pandemic age.* New York: Times Books.

15. Sturm, A., Amino, R., van de Sand, C., Regen, T., Retzlaff, S., Rennenberg, A., Krueger, A., Pollok, J. M., Ménard, R., and Heussler, V. T. (2006). Manipulation of host hepatocytes by the malaria parasite for delivery into liver sinusoids. *Science 313*, 1287–1290.
16. HHMI News. (2006). Movie spies on malaria parasite's sneaky behavior. www.hhmi.org/news/menard20060804.html.
17. da Cruz, F. P., Martin, C., Buchholz, K., Lafuente-Monasterio, M. J., Rodrigues, T., Sonnichsen, B., Moreira, R., Gamo, F. J., Marti, M., Mota, M. M., Hannus, M., and Prudêncio, M. (2012). Drug screen targeted at plasmodium liver stages identifies a potent multistage antimalarial drug. *Journal of Infections Diseases 205*, 1278–1286.
18. The Scientist Staff (2012) Top ten innovations 2011. *The Scientist 1*, 43.
19. Frischknecht, F., Baldacci, P., Martin, B., Zimmer, C., Thiberge, S., Olivo-Marin, J. C., Shorte, S. L., and Ménard, R. (2004). Imaging movement of malaria parasites during transmission by Anopheles mosquitoes. *Cellular Microbiology 6*, 687–694.
20. Angrisano, F., Delves, M. J., Sturm, A., Mollard, V., McFadden, G. I., Sinden, R. E., and Baum, J. (2012). A GFP-actin reporter line to explore microfilament dynamics across the malaria parasite lifecycle. *Molecular and Biochemical Parasitology 182*, 93–96.
21. Angrisano, F., Riglar, D. T., Sturm, A., Volz, J. C., Delves, M. J., Zuccala, E. S., Turnbull, L., Dekiwadia, C., Olshina, M. A., Marapana, D. S., Wong, W., Mollard, V., Bradin, C. H., Tonkin, C. J., Gunning, P. W., Ralph, S. A., Whitchurch, C. B., Sinden, R. E., Cowman, A. F., McFadden, G. I., and Baum, J. (2012). Spatial localisation of actin filaments across developmental stages of the malaria parasite. *Plos One 7*(2), e32188.
22. Frischknecht, F., Martin, B., Thiery, I., Bourgouin, C., and Ménard, R. (2006). Using green fluorescent malaria parasites to screen for permissive vector mosquitoes. *Malaria Journal 5*, 23–30.
23. Marrelli, M. T., Li, C. Y., Rasgon, J. L., and Jacobs-Lorena, M. (2007). Transgenic malaria-resistant mosquitoes have a fitness advantage when feeding on Plasmodium-infected blood. *Proceedings of the National Academy of Sciences of the United States of America 104*, 5580–5583.
24. Corby-Harris, V., Drexler, A., Watkins de Jong, L., Antonova, Y., Pakpour, N., Ziegler, R., Ramberg, F., Lewis, E. E., Brown, J. M., Luckhart, S., and Riehle, M. A. (2010) Activation of *Akt* signaling reduces the prevalence and intensity of malaria parasite infection and lifespan in *Anopheles stephensi* mosquitoes. *PLOS Pathogens 6*, e1001003.
25. Wang, S. B., Ghosh, A. K., Bongio, N., Stebbings, K. A., Lampe, D. J., and Jacobs-Lorena, M. (2012). Fighting malaria with engineered symbiotic bacteria from vector mosquitoes. *Proceedings of the National Academy of Sciences of the United States of America 109*, 12734–12739.
26. Wadman, M. (2011). Sickle-cell mystery solved. *Nature News 479*. doi:10.1038/nature.2011.9342.
27. Cyrklaff, M., Sanchez, C. P., Kilian, N., Bisseye, C., Simpore, J., Frischknecht, F., and Lanzer, M. (2011). Hemoglobins S and C interfere with actin remodeling in Plasmodium falciparum–infected erythrocytes. *Science 334*, 1283–1286.
28. Kakkilaya, B. S. (2012). Malaria site; History of malaria treatment. http://www.malariasite.com/malaria/history_parasite.htm.
29. Enserink, M. (2010). Malaria's drug miracle in danger. *Science 328*, 844–846.

30. Adjalley, S. H., Johnston, G. L., Li, T., Eastman, R. T., Ekland, E. H., Eappen, A. G., Richman, A., Sim, B. K. L., Lee, M. C. S., Hoffman, S. L., and Fidock, D. A. (2011). Quantitative assessment of Plasmodium falciparum sexual development reveals potent transmission-blocking activity by methylene blue. *Proceedings of the National Academy of Sciences of the United States of America 108*, E1214–E1223.
31. Vogel, G. (2010). The "do unto others" malaria vaccine. *Science 328*, 847–848.
32. Butler, D. (2013). Zapped malaria parasite raises vaccine hopes. *Nature News*, doi:10.1038/nature.2013.13536.
33. Seder, R. A., Chang, L.-J., Enama, M. E., Zephir, K. L., Sarwar, U. N., Gordon, I. J., Holman, L. A., James, E. R., Billingsley, P. F., Gunasekera, A., Richman, A., Chakravarty, S., Manoj, A., Velmurugan, S., Li, M., Ruben, A. J., Li, T., Eappen, A. G., Stafford, R. E., Plummer, S. H., Hendel, C. S., Novik, L., Costner, P. J. M., Mendoza, F. H., Saunders, J. G., Nason, M. C., Richardson, J. H., Murphy, J., Davidson, S. A., Richie, T. L., Sedegah, M., Sutamihardja, A., Fahle, G. A., Lyke, K. E., Laurens, M. B., Roederer, M., Tewari, K., Epstein, J. E., Sim, B. K. L., Ledgerwood, J. E., Graham, B. S., Hoffman, S. L., and VRC 312 Study Team (2013). Protection against malaria by intravenous immunization with a nonreplicating sporozoite vaccine. *Science 341*, 1359–1365.
34. Baird, C., and Cann, M. (2008). *Environmental chemistry*. 4th ed. New York: W. H. Freeman.
35. Riehle, M. M., Guelbeogo, W. M., Gneme, A., Eiglmeier, K., Holm, I., Bischoff, E., Garnier, T., Snyder, G. M., Li, X. Z., Markianos, K., Sagnon, N., and Vernick, K. D. (2011). A cryptic subgroup of Anopheles gambiae is highly susceptible to human malaria parasites. *Science 331*, 596–598.
36. Enserink, M. (2002). What mosquitoes want: Secrets of host attraction. *Science 298*, 90–92.
37. Turner, S. L., Li, N., Guda, T., Githure, J., Carde, R. T., and Ray, A. (2011). Ultra-prolonged activation of CO_2-sensing neurons disorients mosquitoes. *Nature 474*, 87–91.
38. Turner, S. L., and Ray, A. (2009). Modification of CO_2 avoidance behaviour in Drosophila by inhibitory odorants. *Nature 461*, 277–281.
39. Stopfer, M. (2011). Malaria mosquitoes bamboozled. *Nature 474*, 40–41.
40. Windbichler, N., Menichelli, M., Papathanos, P. A., Thyme, S. B., Li, H., Ulge, U. Y., Hovde, B. T., Baker, D., Monnat, R. J., Burt, A., and Crisanti, A. (2011). A synthetic homing endonuclease-based gene drive system in the human malaria mosquito. *Nature 473*, 212–215.
41. http://news.sciencemag.org/biology/2011/04/disease-proof-mosquito-could-spread-wildfire
42. Fang, W. G., Vega-Rodriguez, J., Ghosh, A. K., Jacobs-Lorena, M., Kang, A., and St. Leger, R. J. (2011). Development of transgenic fungi that kill human malaria parasites in mosquitoes. *Science 331*, 1074–1077.
43. Catteruccia, F., Crisanti, A., and Wimmer, E. A. (2009). Transgenic technologies to induce sterility. *Malaria Journal 8*, S7–15.
44. Catteruccia, F., Benton, J. P., and Crisanti, A. (2005). An Anopheles transgenic sexing strain for vector control. *Nature Biotechnology 23*, 1414–1417.
45. Walters, M., Morrison, N. I., Claus, J., Tang, G., Phillips, C. E., Young, R., Zink, R. T., and Alphey, L. (2012) Field longevity of a fluorescent protein marker in an engineered strain of the pink bollworm, Pectinophora gossypiella. *PLOS One 7*, e38547.

CHAPTER 5

Dengue Fever

Year	Countries with Severe Dengue Epidemics
1970	9
2011	127

It is a virus, carried by a mosquito, and begins as a headache of such voltage that you tremble and can't stand or sit. You're knocked flat; your muscles ache, you're doubled up with cramp and your temperature stays over a hundred. Then your skin becomes paper-thin, sensitive to the slightest touch—the weight of a sheet can cause pain. And your hair falls out—not all of it, but enough to fill a comb. These severe irritations produce another agony, a depression so black the dengue sufferer continually sobs. All the while your bones ache, as if every inch of you has been smashed with a hammer.

Paul Theroux, *"Dengue Fever"*

In late 1919, local industries and the city council of Perry, Florida, sent George Simons, a sanitary engineer for the state, an emergency request to control the malaria mosquitoes in the area. Simons went to work, and in 6 months his team cleared 15 acres of land, dynamited an old dam, opened 3 miles of creek beds, and constructed three new bridges. The project was the biggest antimosquito campaign conducted in the United States. It was an unqualified success, and in just 1 year the malaria infection rate in Perry dropped by 90 percent. Simons tried to expand his antimosquito plans, but he couldn't find any funding for the project because no one was interested in mosquito control. That changed after a dengue outbreak swept through the southeastern United States in 1922, infecting more than 200,000 people. Simons called the epidemic a "blessing" and said, "It gave us an admirable chance to awaken interest in the whole program and start a statewide control program. The people were in a receptive mood to consider mosquito control measures, and today there isn't a community in Florida that wants to tolerate another dengue outbreak." (1, p. 112) The outbreak also opened the door for other antimosquito crusaders around the United States. They fought a hard war against mosquitoes using ditching and drainage as their principal weapons, and that war was about to change. In June 1943, DDT was sprayed from an airplane for the first time. This technique would prove so effective at killing mosquitoes that by 1946 the Communicable Disease Center had sprayed more than 1 million homes with DDT, and within 3 years of initiating DDT spraying in Florida, dengue and malaria were eliminated in that state.

However, dengue fever returned to Florida in 2009. Dengue fever is a re-emerging disease not just in Florida but throughout the world. It once was thought to be under control, but in the last 50 years, its incidence has increased 30-fold worldwide, and it is now the fastest-spreading infectious disease. According to the Pan American Health Organization, the number of dengue cases in the Western Hemisphere doubled from 2012 to 2013. DDT was responsible for eliminating dengue fever from Europe and the United States, as well as for its decline in the tropics. However, due to DDT's persistence and impact on bird populations, its use has been discontinued in most countries, and applications are now permitted only to control pests carrying a life-threatening disease, such as malaria. Although 40 percent of the population of the world is threatened by dengue fever and there is no treatment or vaccine for the disease, DDT applications to control dengue fever are prohibited. Having dengue fever is a painful, miserable experience, but it is not deadly enough to warrant the use of DDT.

Today dengue fever is 20 times more common than the flu. Consequently, many researchers are trying to find a cure or a vaccine for dengue fever, but there seems to be no medicinal solution on the horizon. This has led many entomologists to look for ways to control the mosquitoes that transmit dengue fever without having to rely exclusively on pesticides. Three of the research groups working in the area of mosquito control have attracted a large amount of media attention.

A group led by Scott O'Neill, a biologist at Monash University in Australia, has infected mosquitoes with bacteria that will prevent the dengue viruses from taking residence in the mosquito. They are using fluorescent proteins to maximize the number of dengue-fighting bacteria per mosquito. O'Neill chose to release the insects close to home. Dengue outbreaks are less common in Queensland, Australia, than in many less developed tropical countries, but the group chose Australia because

O'Neill did not want to be accused of taking advantage of another country's misfortunes. He hopes that successful field tests at home will result in invitations to come and repeat them in foreign areas infested with dengue. And it has worked. According to O'Neill, "We were getting bombarded by people around the world, from different governments, wanting us to come work in their countries because people are so desperate for something to try and stop dengue." (2)

An American research group has been testing genetically modified male mosquitoes that have been designed to mate with native mosquitoes and produce flightless female offspring in large mosquito cages in Mexico. They bought the land they are using for their experiments and have been consulting with local inhabitants in preparation for a field test of their mosquitoes.

Finally, a British biotech company, Oxitec, has already released millions of genetically modified mosquitoes in trials in the Grand Cayman Islands, Brazil, and Malaysia. These mosquitoes express fluorescent proteins so that they can be distinguished from normal mosquitoes (figure 5.1). From 2002, when the company made the first transgenic mosquito, Oxitec's researchers have been using fluorescent markers. First they used mosquitoes that expressed the fluorescent proteins all over, but currently they often use nuclear localization so the fluorescent proteins are located in defined spots; this makes them much easier to distinguish from background autofluorescence, which is evenly spread

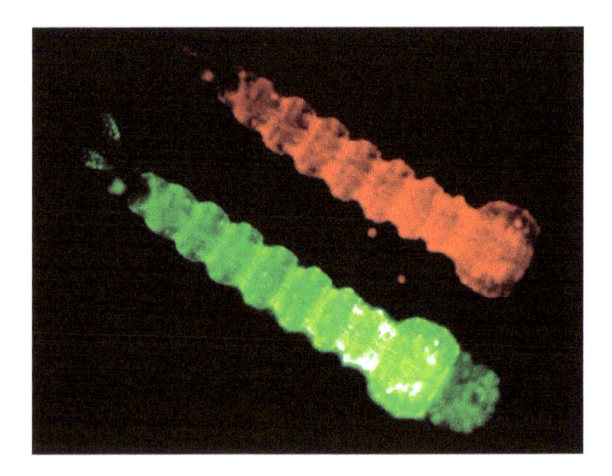

Figure 5.1 *A. aegypti* transmit dengue fever. A small biotech company, Oxitec, has genetically modified these mosquitoes so that they produce offspring that do not progress beyond the larval stage. The genetically modified mosquitoes also contain fluorescent proteins so that they can be distinguished from the wild mosquitoes in field trials. (Courtesy of Derric Nimmo/Oxitec Ltd.)

across the body. They also use fluorescent sperm to discover whether their transgenic mosquitoes are fertilizing wild-type mosquitoes. On the subject of fluorescent proteins, Luke Alphey, the chief scientific officer of Oxitec, says, "We use it in everything we do, but in an indirect way to mark and track insects. The core of what we do is to make genetically sterile insects that have a lethal gene so that if the male carries copies of the gene its offspring will inherit it and will die. Every step of the way we need to recognize the transgenic in a much larger pool of non-transgenics. It is important to know whether our transgenic males are mating with wild females. We can collect offspring and see whether they are fluorescent or not." (Personal interview.)

In chapter 4 we saw that genetically modified *Anopheles* with glowing gonads have been created to separate male and female malaria mosquitoes for sterile insect technique experiments. However, these mosquitoes have never been released. The Oxitec mosquitoes are the first genetically modified insects ever released in nature. Before delving into the controversy about the release of self-destructing genetically modified mosquitoes, it is instructive to know more about dengue fever and the mosquitoes responsible for transmitting it.

There are some similarities between malaria and dengue fever: both are transmitted by mosquitoes, are found in the tropics and subtropics, and cause an unpleasant fever. However, there are more differences than similarities. A virus, not a parasite, causes dengue fever; the incidence of dengue fever is increasing; and mosquitoes from the genus *Aedes*, and not *Anopheles,* transmit it.

Aedes aegypti, also known as the yellow fever mosquito, is the most common transmission vector of dengue fever, and although it prefers humans, it will feed on the blood of nonhumans when it is very hungry. Its favored feeding sites are around the ankle area, and it feeds for 1- to 2-hour periods in the morning and late afternoon. *Aedes* are smaller and quieter than the mosquitoes typically found in the United States. And although they originated in Africa, they are now found in tropical and subtropical areas throughout the world. It is these stealthy bloodsuckers that are brazen enough to attack humans in broad daylight that we need to either eradicate or make inhospitable to the dengue viruses.

The dengue viruses inhabit the cells lining the gut of the *Aedes* mosquito. After an incubation period of 8 to 10 days, the mosquito is capable of transmitting the viruses to a human or a nonhuman primate during a blood meal. The viruses do not harm the mosquito, but once a dengue virus infects a mosquito, the insect is infected for life. The virus can also be transmitted to subsequent mosquito generations through the eggs the female lays. If an infected mosquito bites a person, the virus enters the bloodstream with the mosquito's saliva and anticoagulant. Immediately after entering the human, it replicates in the lymph nodes before being disseminated through the blood to various tissues in the body. In the blood, it binds and enters the white blood cells and causes fevers, headaches, and muscle and joint pains. In a few extreme cases, leakage of the blood plasma through the walls of the small blood vessels into the body cavity occurs, resulting in bleeding. This is known as *dengue hemorrhagic fever* and causes very low blood pressure, which is called *dengue shock syndrome*. The dengue viruses are much smaller and simpler than the malaria parasite, and their size and simplicity are their defense against the human immune system. The virus has no need to go through a complex life cycle. It circulates in the human bloodstream for 2 to 7 days, during which time other *Aedes* mosquitoes feeding on the subject can be infected with the virus.

Viruses are made of only two components, genetic material—either DNA or RNA—and a protein envelope. In order to reproduce, they have to infect some cell-based organism and introduce their genetic material into their host. Viruses can be found in all forms of life; they can even infect bacteria and parasites. This makes them the most common form of life on earth, although there is some argument whether viruses are alive or not. Viruses have very few genes and have the highest mutation rates of any organism. They can also exchange genetic information with other viruses. When two different viruses infect the same cell, they can form a new virus that contains parts from one virus and completely different parts from another. This genetic mixing allows them to evade immune systems, avoid drugs, and successfully invade new species.

In 1907, it was first shown that a virus is responsible for dengue fever. At the time, yellow fever was the only other disease that was known to be caused by a virus. The dengue viruses are flaviviruses. In Latin, *flavus* means "yellow"; this indicates that the dengue viruses are a member of the yellow fever–like viruses, whose other members are Japanese encephalitis, West Nile encephalitis, tick-borne encephalitis, and, of course, yellow fever itself.

The first descriptions of patients with symptoms approximating dengue fever were reported in a Chinese medical dictionary as early as A.D. 400. Fever, pain in the eyeballs, a rash, as well as oral, vaginal, and intestinal bleeding were some of the symptoms used to characterize the disease. Numerous outbreaks of dengue fever were described in the seventeenth and eighteenth centuries. The disease struck Philadelphia in 1780, when all the residents living along the Delaware River waterfront were afflicted.

They called the disease "break-bone fever," a name still associated with the disease today. In 1827–1828, the disease ravaged St. Thomas and St. Croix, where the joint pain associated with fever caused those suffering from it to walk with a dandified gait, leading to another name for the disease, dandy fever. The currently used name, dengue fever, has similar origins. It comes from a 1801 outbreak in Madrid in which patients were described as having a fastidious or careful walk (*dengue* is Spanish for "fastidious"). (3)

Dengue hemorrhagic fever and dengue shock syndrome, two related conditions, have only appeared much more recently. The first clinical descriptions were reported in dengue outbreaks in Australia in 1897 and Greece in 1928. According to the Centers for Disease Control and Prevention, "The principal symptoms of dengue fever are high fever, severe headache, severe pain behind the eyes, joint pain, muscle and bone pain, rash, and mild bleeding (e.g., nose or gums bleed, easy bruising). Generally, younger children and those with their first dengue infection have a milder illness than older children and adults. Dengue hemorrhagic fever is characterized by a fever that lasts from 2 to 7 days, with general signs and symptoms consistent with dengue fever. When the fever declines, symptoms including persistent vomiting, severe abdominal pain, and difficulty breathing, may develop. This marks the beginning of a 24- to 48-hour period when the smallest blood vessels (capillaries) become excessively permeable ('leaky'). . . . This may lead to failure of the circulatory system and shock, followed by death, if circulatory failure is not corrected. In addition, the patient with dengue hemorrhagic fever has a low platelet count and hemorrhagic manifestations, tendency to bruise easily or other types of skin hemorrhages, bleeding nose or gums, and possibly internal bleeding." (4)

The dengue viruses are spherical and have a diameter of 40 to 50 nanometers. There are four strains, also known as *serotypes*, of the virus—DENV-1, DENV-2, DENV-3, and DENV-4—that all evolved from a common ancestor that inhabited a nonhuman primate and independently entered humans. Cases of dengue hemorrhagic and dengue shock fever have only been reported for patients who have had an initial case of dengue fever, followed by a subsequent infection by a strain of dengue that didn't cause the initial infection. The longer the interval between infections with different dengue serotypes, the more severe the symptoms. (5)

The RNA of the dengue viruses has been sequenced. It is made up of 11,000 nucleotide bases that code for three proteins that make up the virus particle and seven other proteins that are made in the infected host cells and are responsible for viral replication. All four serotypes of the dengue fever virus share only about 65 percent of the genome, yet they are responsible for nearly identical symptoms. DENV-2 viruses that express GFP or firefly luciferase have been created and can be followed in live mice. (6)

Somehow, the simple dengue viruses with RNA coding for just 10 proteins can change the production of 147 different proteins expressed by *A. aegypti*. It makes the mosquito hungrier for human blood and its saliva more hospitable to the viruses, and it changes the protein mix in the antennae of the mosquitoes, making them more sensitive to odors and thereby increasing the mosquito's ability to find a victim. (7)

The number and severity of dengue infections have been escalating since the Second World War, culminating in a 30-fold increase between 1960 and 2010. The most dramatic recent increases have occurred in the Caribbean and Central America. According to the World Health Organization, "Not only is the number of cases increasing as the disease spreads to new areas, but explosive outbreaks are occurring." (8) Currently about 50 to 100 million people are infected in more than 100 countries. About 40 percent of the world's population is at risk of dengue fever infection, making it the most common mosquito-borne disease.

A. aegypti, the vector for the dengue viruses, thrives in urban environments; it is more at home in the city than in the jungle. The increase in dengue infections can be attributed to increased population growth, urbanization, and global warming. *A. aegypti* prefers to lay its eggs in man-made containers such as metal drums and earthenware jars, as well as in refuse like plastic containers and abandoned car tires that are capable of collecting rainwater. They live indoors and reside in darkly lit closets and cupboards. They are stealthy stingers; there is no annoying buzzing for these silent mosquitoes, which can bite up to 20 people a day. Controlling and limiting mosquito habitats is extremely difficult because mosquitoes can lay eggs in a single drop of water, and *A. aegypti*'s ideal habitat is close to human homes, where it avoids the spraying of pesticides. This has led to severe restrictions in several countries: in Singapore, a homeowner can be fined for having *Aedes* breeding sites, such as a glass of water, in the garden; in Malaysia, the home of anyone who contracts dengue fever and those of their neighbors has to be sprayed with pesticides both inside and outside.

A. aegypti is also known as the yellow fever mosquito because it is also the carrier of the yellow fever virus. From 1947 to 1965, a concerted eradication of *A. aegypti* was conducted to control the spread of yellow fever. It was very successful, but after the cessation of the *Aedes* eradication programs due to bans on the use of DDT and the creation of an effective yellow fever vaccine in 1937, there has been a worldwide resurgence of

A. aegypti and dengue fever, making it the fastest-spreading infectious disease in the world.

In 2009, there were 27 reported dengue cases in Key West, Florida, and in 2010 there were 66 cases. These were the first recorded dengue episodes in southern Florida in more than 60 years. The last major Floridian dengue outbreak, with approximately 2,000 cases, occurred in 1934. Amid worries that the recent dengue outbreaks might seriously affect tourism in the area, a new executive director for the Florida Keys Mosquito Control District was hired: Michael Doyle. There are 44 mosquito species in the Keys, and *A. aegypti* are the only ones capable of transmitting the dengue viruses. Doyle was given a $10 million annual budget, four helicopters, and two airplanes to ensure that there would be no more dengue outbreaks in the Florida Keys. According to Doyle, "The dengue cases were a big deal. It was the first time [the disease] had been back in more than 60 years. The concern is that the Keys could be a way for dengue to get a new foothold, or a refoothold, in the United States." (9, p. 15) This is not a far-fetched concern, as *Aedes* mosquitoes are found all the way up the Florida coast to Miami, and local health authorities are concerned that the virus may make its way up north to more populated areas and big cities as it did in 2013, when dengue fever was diagnosed in Miami-Dade patients who had not left the city itself and must have been bitten by local mosquitoes.

According to some estimates, there are about 40,000 mosquitoes per person in the United States. Most parts of the country are warm enough for the mosquitoes to survive for only 4 months. In Key West, however, the mosquito season lasts for 10, and sometimes 12, months a year. Before the advent of DDT, this was a major problem. According to Chris Sweeney of the *Miami New Times*, "In the late 19th century, when Florida was referred to as 'the Devil's property,' swarms were so dense in some areas that it was impossible to breathe without inhaling mouthfuls of mosquitoes." (9, p. 15)

This is the root of Michael Doyle's nightmares. His solution is as dramatic as it is controversial; he wants Key West, home to many laid-back Jimmy Buffet fans and environmentalists, to be a test site for the first US release of genetically modified self-destructing *Aedes* mosquitoes.

Aedes are also found in Arizona, where there have been no dengue fever outbreaks as of yet, while nearby Texas hospitals and doctors have been used to treating patients with dengue fever for the last 30 years because they have had regular small, confirmed dengue fever outbreaks. Michael Riehle, a mosquito entomologist at the University of Arizona, thinks that the less hospitable climate in Arizona prevents *Aedes* from living long enough to spread the dengue viruses. Mosquitoes live 2 to 3 weeks and have to be at least 14 days old before they can pass the dengue viruses on during a blood meal. Riehle is having some problems proving his hypothesis because there is no way to determine how old a wild mosquito is. He is looking for genes that are turned on or off after the mosquito is 2 weeks old. These genes could be tagged with fluorescent proteins and used to determine the age of the mosquitoes. Riehle is also hoping to modify the genes in order to breed perfectly healthy *A. aegypti* that can compete with their wild-type brethren but die young, so that even if they are infected they cannot transmit the virus. (*10*)

Zach Adelman from the Department of Entomology at Virginia Tech also studies *Aedes* mosquitoes. As a graduate student he decided to start working with mosquitoes, mainly because he saw that there was much research that needed to be done, and because he did not want to work with mouse models. He is happy with his decision, even if he occasionally has mosquito strains that don't like the artificial mosquito feeder and prefer human blood. To accommodate them, he says, "I put a seat down, roll-up my pant leg and throw my leg over the cage. My calf is the easiest spot I have found. It's where I won't scratch the bites, and I don't have to look at them either, that's important because I have to keep leg there for twenty minutes to make sure that they all feed. I can sit there reading a book, or a paper. Fortunately, I don't react too strongly to the bites." (*11*)

Adelman wonders why *Aedes* are the only mosquito genus to transmit the dengue viruses. Invertebrates, such as mosquitoes, use something called the *RNA interference pathway* to protect themselves from viral infections. Adelman suspects that dengue viruses can suppress or otherwise evade the RNA interference pathway in *Aedes* mosquitoes, thereby neutralizing their antiviral defenses. To identify the active viral suppressors, as well as mosquito genes important in antiviral defense, Adelman has created genetically modified *Aedes* that have the genes for both red and green fluorescent proteins in their eyes. The red fluorescent protein is continuously expressed, but the green one is expressed only when the RNA interference pathway is suppressed. All Adelman's transgenic mosquitoes have red fluorescent proteins in their eyes, and only those with compromised antiviral defense systems have the green fluorescent protein (figure 5.2). The *Aedes* Adelman uses in these experiments are a special strain that have been passed around mosquito research groups since the 1960s. They have white eyes, which makes the green and red fluorescence easily detectable. Naturally occurring *Aedes* have black pigments in their eyes that would absorb most of the fluorescence. Having created the mosquitoes, Adelman has been using them to examine their

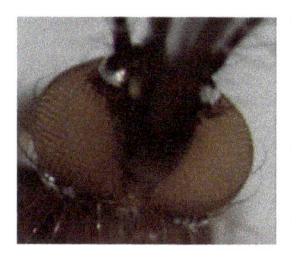

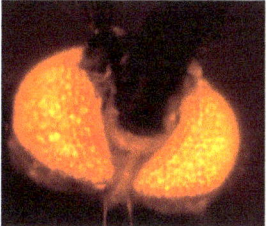

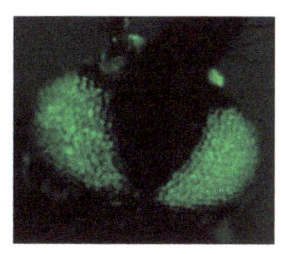

Figure 5.2 *A. aegypti* designed to determine whether the dengue viruses are able to compromise the mosquito's antiviral defenses. All transgenic mosquitoes have red fluorescent eyes to differentiate them from the wild mosquitoes, and those with compromised RNA interference pathways fluoresce green. The RNA interference pathway, which has a major responsibility for the antiviral defense of the mosquito, has been inhibited in the *Aedes* shown here. No fluorescence is seen under visible light (left); using blue light and the right set of filters, the red (center) and green (right) fluorescence in the eye can be imaged. (*13*)

antiviral defense system. It has been a trying time for Adelman, and after 4 years he has concluded that the presence of the infecting viruses is causing some complexities in mosquitoes' antiviral defense that he has yet to unravel. "Right now, it's a black box," says Adelman. "We know how things work only in general and very vague terms. If we find out more about what's going on, we could have a better idea of how to implement some kind of control. Maybe there are 10 genes involved or 100. We just don't know. We know of a couple of genes so far. We're just trying to find more, so we can understand how mosquitoes defend themselves against these viruses. Only then will we be able to understand what turns a simple pest into a potential killer." (*12, 13*) The dengue virus and its relations with its mosquito host are clearly complicated, and although it is very important to understand why the *Aedes* mosquitoes are immune to the dengue virus–associated pathology, there might be easier ways to control the spread of dengue fever.

The rapid spread of dengue fever has led to an associated increase in research. From 2002 to 2011, more than 8,000 dengue research papers were published, a threefold increase over the previous decade. Dengue fever is now the fastest-growing research area in tropical diseases. (*14*) Despite all the research occurring in the field, there is no treatment for dengue fever. At best doctors can give their patients supportive care, such as painkillers and liquids to keep them hydrated.

Fortunately, most cases are not life-threatening. The mortality of dengue fever is between 1 and 5 percent when patients do not undergo medical treatment. It is rare for dengue fever to progress to the more severe forms, and the severe forms of the disease can be managed by early intravenous rehydration. During the early stages of dengue there is no way to predict whether patients will progress to one of the two severe forms. Later in the infection, warning signs are bleeding, severe and continuous stomach pains, persistent vomiting, large temperature changes, and fainting, which can indicate the start of blood plasma leakage. Early supportive treatment at this point can prevent the progression of the disease to massive hemorrhaging from the eyes, nose, mouth, and vagina and multiple organ and respiratory failure, and can reduce mortality rates for patients with dengue hemorrhagic fever from 20 percent to 1 percent.

Although dengue vaccines have been under development for several decades, the likelihood of a licensed vaccine hitting the market in the next 10 years is slim. The main difficulty in designing a vaccine is that it must be effective against all four strains of the virus. Recent reports have documented cases of dengue hemorrhagic fever and dengue shock syndrome that have occurred 20 years after the initial infection. Because there is no vaccine or treatment for dengue fever, when outbreaks occur, they can only be controlled by reducing the number of *Aedes* mosquitoes in the affected area or preventing the mosquitoes from carrying the pathogen.

The first and most obvious step in controlling the numbers of *Aedes* during a dengue fever outbreak is to limit the insect's breeding areas by reducing the number of sites with stagnant water. If the standing water cannot be removed, then pesticides and organisms that feed on mosquito larva—guppies and copepods are favorites—are added to the water. However, the use of guppies and copepods is limited because they frequently need to be reintroduced in some habitats, and their mass rearing requires expertise and facilities that are often not present in affected areas. *A. aegypti*'s desire to live in close contact with humans means that water used for domestic use also needs to be treated. This includes drinking water, and thus the larva-killing pesticides cannot have toxic effects on humans and must be tasteless, which severely limits the number of larvicides available. Because *A. aegypti* feed during the day, mosquito netting is also not practical. Indoor spraying with DDT is effective, but the human health consequences of dengue fever are not severe enough to warrant the use of the pesticide.

Yorkeys Knob and Gordonvale, two small towns in Queensland, Australia, had sporadic outbreaks of dengue fever prior to 2011, but now the dengue virus has lost its ride in these two towns. Although the numbers of *A. aegypti* and the frequency of their human blood meals remain unchanged in Yorkeys Knob and Gordonvale, the mosquitoes in these towns can no longer carry dengue viruses. They no longer transmit dengue viruses, and dengue fever is hopefully no more than an unpleasant memory in these two towns. The mosquitoes were made resistant to dengue viruses by infecting them with a bacterium, *Wolbachia*, that is commonly found in insects.

The *Wolbachia*-infected mosquitoes are the result of an idea that Scott O'Neill, working in Australia, had in the 1990s He knew that *Wolbachia*-infected fruit flies would not transmit any RNA virus, and he thought *Wolbachia*-infected *A. aegypti* would act in the same way and not transmit dengue, an RNA virus. However, even though *Wolbachia* infections are common in many insects, he couldn't infect sufficient numbers of *Aedes* with the bacteria. He explains why he persisted: "I thought the idea was a good idea, and I don't think you get too many ideas in your life, actually. At least I don't. I'm not smart enough. So I thought this idea was a really good idea." (*15*)

It wasn't easy, but by obsessively trying new and different ways to infect *Aedes* with a strain of the bacteria wMel, obtained from the fruit fly *Drosophilia melanogaster*, O'Neill and his colleagues finally managed to overcome the mosquitoes' resistance to *Wolbachia*. The wMel strain of *Wolbachia* severely compromises dengue virus infections in *A. aegypti*. Because the results with caged mosquitoes were so positive, the *Wolbachia* infection rapidly spread through the caged mosquito population, and *Wolbachia*-infected mosquitoes were resistant to the dengue viruses, the researchers from Monash and Queensland Universities started designing field experiments. After consultation with the Australian government, all parties involved agreed to regulate the release of the *Wolbachia*-infected *A. aegypti* by using existing legislation as a veterinary chemical product under the supervision of the Australian Pesticides and Veterinary Medicines Authority. For 2 years, before the release of about 300,000 infected mosquitoes, the communities of Yorkeys Knob and Gordonvale were informed of the experiment and were extensively consulted. All residents were asked to give written permission for "three separate activities to take place on their properties. These included pre-release suppression of mosquitoes by removing water from potential breeding sites, release of mosquitoes containing *Wolbachia* and the placement of monitoring traps in resident yards."

Seventeen inhabitants of Yorkeys Knob, population 907, elected not to participate in the program, and 11 of the 1,207 inhabitants of Gordonvale also declined to participate in the study. Infected mosquitoes were released only on properties of residents who gave permission, and no mosquitoes were released on the properties, or the adjacent properties, of households that had chosen not to participate in the study. Over 2.5 months, roughly 150,000 infected mosquitoes were released. The experiment was a success. One hundred percent of all mosquitoes caught in the monitoring traps in Yorkeys Knob were infected by the *Wolbachia* bacteria, and more than 80 percent of the Gordonvale mosquitoes were infected and were presumably unable to transmit the dengue virus. Infected mosquitoes were even found several kilometers away from the release sites, including in a suburb of Gordonvale that is separated from the release sites by a major highway. Highways are considered a major movement barrier to mosquitoes and mosquito gene flow. The infected mosquitoes were not expected to spread far from their release point because *A. aegypti* are known to be weak dispersers. They move less than 400 meters in their lifetimes unless they can hitch a ride on their victim's clothing or in a car.

It will be very interesting to see if denying the dengue viruses their access to the *Aedes* mosquito will eradicate dengue fever permanently from Yorkeys Knob and Gordonvale. O'Neill and his co-workers are optimistic, "This should provide a strategy for sustainable dengue control at low cost, with a relatively simple deployment system suitable for implementation in developing countries. The next step is a disease endpoint trial to test efficacy of the method for dengue and dengue haemorrhagic fever control, ongoing monitoring in and around the release area to test for persistence, and releases to test the spatial spread of the infection across a populated area." (*16*, p.457)

Twenty years of persistent research allowed Scott O'Neill and his students to infect mosquitoes in the lab with *Wolbachia*, which were less likely to harbor the dengue viruses than the *Wolbachia*-free mosquitoes. They then did limited field tests, which proved that the infected mosquito could spread the bacterial infection to the native mosquito population. The work was published in two back-to-back papers in *Nature* during 2011. This would be the academic high point of almost any scientist's career, but O'Neill wants more; he hopes his *Wolbachia* will infect all the *Aedes* in the world and make them inhospitable to the dengue viruses. He has already extended his field release to Babinda, Australia's wettest town, where he has once again shown that his *Wolbachia*-infected mosquitoes can pass on their *Wolbachia* infection to the local *Aedes*. Now he is dreaming of worldwide dengue domination, with collaborators in Vietnam, Brazil, Thailand, and the United States preparing for measured releases of *Wolbachia*-infected *Aedes*. (*16, 17*)

Wolbachia infect 65 percent of all insect species and are known to manipulate their host's reproductive systems, yet little is known about the molecular interactions between the *Wolbachia* bacterium and its host. At the same time as the field tests are being conducted, the Queensland researchers are continuing their investigations of *Wolbachia* in the laboratory. In order to establish how the bacteria change the mosquito's intracellular environment to ensure their own survival, Scott O'Neill and his coworkers have gone searching for proteins whose expression is increased when the *Aedes* is infected with *Wolbachia*. Using GFP, they showed that *Wolbachia*-infected mosquitoes made more metalloprotease than uninfected ones and that inhibiting metalloprotease expression resulted in significant decreases in *Wolbachia* infection. Now, by increasing *Aedes* metalloprotease expression, they can load up *Aedes* with even more *Wolbachia* to ensure that there is zero chance that they can transmit the dengue viruses. (18)

In the Grand Cayman Islands, about as far from Australia as you can possibly get, 3.3 million genetically modified *A. aegypti* were released by Oxitec, a small British company. The Australian study was the first research project in which deliberately infected mosquitoes were released into the wild, while the Grand Cayman study was the first field trial with genetically modified insects.

The Oxitec trials are based on work that their chief scientific officer, Luke Alphey, published in the 1990s while he was at Oxford University. He found a gene that kills off all the *Aedes* offspring when they are in the larval stage and also found a way to suppress the expression of the gene with the antibiotic tetracycline. This means that in the presence of tetracycline the larvae develop normally, allowing researchers to grow large batches of transgenic *A. aegypti*, while in the wild, where the mosquitoes will have no contact with the antibiotic, the larvae of the genetically modified mosquitoes will die.

At a meeting of the American Society of Tropical Medicine and Hygiene where Alphey was presenting his work, he was approached by authorities from the Grand Cayman Islands who were interested in hosting field trials of Oxitec's self-destructing mosquitoes. Although the Grand Caymans have had no reported cases of dengue fever, *A. aegypti* is very common there, and authorities felt it was just a question of time before dengue made its way onto their islands; thus, they wanted to take action to preempt the virus. Oxitec was ready to move out of the laboratory and was were excited by the collaboration, particularly because the *A. aegypti* on the island exhibit high levels of insecticide resistance, demonstrating the need for alternative methods of mosquito control.

In a small initial trial used to establish how many transgenic males were required per female, 16,000 transgenic males were released on the island. Because the males are essentially sterile, the Oxitec trials are a variation of the sterile insect technique described in chapter 4 in the discussion of malaria. Two problems have prevented the adoption of sterile insect techniques to control malaria: first, sterilizing male *Anopheles* with radiation, the most common method, reduces their ability to compete for female mates; second, *Anopheles* male and female larvae are difficult to distinguish. Using Oxitec's transgenic mosquitoes, these problems are overcome. There is no need to sterilize the males because genetically modifying them so that all the larvae they produce die has the same effect as sterilization. Furthermore, no glowing green testicles were required to separate the male and female mosquito larva because male *A. aegypti* larva are smaller than their female counterparts and can easily be distinguished. In all field tests, starting with the ones in the Grand Cayman Islands, no genetically modified females were released. This is critical because Oxitec has to prevent the genetically modified insects from breeding with each other in the wild and to ensure that transgenic mosquitoes do not bite any humans—remember, only the females bite. The genetically modified mosquitoes with the self-destruct gene were also designed to express red fluorescent proteins so that the transgenic larvae could easily be distinguished from their wild counterparts (see figure 5.1).

The first big question Alphey and Oxitec wanted answered in the Grand Caymans was whether their genetically modified *Aedes* mosquitoes would mate. The mosquitoes did great in the lab, but they had been reared in the lab for at least 100 generations. According to Alphey, "The males primarily live in 30cm cube cages and they don't have to go anymore than 30cm to find a female and find food. They have never been exposed to rain, they have never been exposed to sunlight, and have never seen predators. The big question was could they find and court the wild females." (personal interview) The fluorescent males were released, and ovitraps were placed around the release sites. The ovitraps are just black water containing cups with a stick in them that attracts the female mosquitoes to lay their eggs. In Alphey's opinion, one of the critical points in Oxitec's history was when its researchers found the first fluorescent pupa in one of the ovitraps. A wild-type female had been impregnated by a genetically modified male! What is more, it wasn't an isolated event; soon the researchers discovered that the transgenic mosquitoes regularly successfully mated with the females and fertilized their eggs. However, they were not as efficient as the wild-type males. In the area where the transgenic males were released, the males made up 16 percent of the male mosquito population, but only 9.6 percent of all the larva were fluorescent nonviable mosquitoes. (*19*)

This was not a surprising finding. Lab-bred male mosquitoes struggle to compete with the wild-type males, and the Oxitec males were in fact more efficient breeders than typical lab-reared males. There may be a number of reasons for the poor showing of males bred in captivity. In some cases it has been observed that lab-raised males will ignore wild females, preferring to mate with other lab-reared mosquitoes. However, this was not a problem in the Grand Cayman studies because only lab-bred males were released. They had no choice and had to find wild-type females. The most likely cause is that the transgenic males may be smaller and less competitive than their wild compatriots because they have been raised in large, dense groups where they don't have to go hunting far and wide for a female companion.

To overcome the fact that lab-reared mosquitoes are not the Lotharios the Oxitec researchers would like them to be, they calculated that they would require a sustained release between 4 and 12 times the release rate used in the smaller proof of concept experiment.

In the Grand Cayman trial, around 3.3 million lab-bred male *A. aegypti* with the inserted gene were released, which resulted in a ratio of about 10 transgenic males for one wild-type male mosquito. With numbers like these, the wild-type females were swamped with genetically modified males, and the odds were high that they would mate with a transgenic male. This full-scale release was successful, and 80 percent of the mosquito population was eradicated.

Following its Grand Cayman releases in 2008, Oxitec has also started field tests with transgenic mosquitoes in Malaysia in 2010 and northern Brazil in 2011. The Brazilian releases took place in Itaberaba, a densely populated suburb in Bahia state, where they were done in conjunction with the Brazilian Ministry of Health and researchers from the University of São Paolo. Before the mosquitoes were released, the local residents were informed of the project and its potential benefits. The public relations campaign, which includes the following Brazilian ditty, must have been a success, because the children of Itaberaba follow the Oxitec truck and run through the releases of the nonbiting self-destructing Oxitec mosquitoes.

> Let him into your house
> He is the solution
> He fights dengue
> And he won't bite anyone
> Protect your health
> He's a good mosquito

Margareth Capurro of the University of São Paulo, who is leading the project, said: "After a long period of contained evaluation work, we started a series of releases in Brazil in February 2011 in the outdoor environment. Then, from December 2011 we commenced a suppression trial and showed that, in the area where we were releasing the sterile male mosquitoes, we could control the mosquito that spreads dengue fever. This was done in a suburb of Juazerio, Bahia state, where mosquitoes are at a very high level all year round. When we started the trial, we were seeing *Aedes aegypti* in about half of the traps we set in and around people's homes. Now we see hardly any. Comparing the area of release to the adjacent area where no releases were made, we have reduced the population of *Aedes aegypti* by 85 percent. We are very excited by the results." (20) The release was so successful that on July 7, 2012, Brazil's minister of health, Alexandre Padilha, officially opened a new facility to create enough mosquitoes to protect a town of approximately 50,000 inhabitants from *A. aegypti*. (20) At maximum production the facility will produce 4 million sterile mosquitoes a week. In the first 6 months of 2012, Brazil reported 431,200 dengue cases, so it is not surprising that the Brazilian Health authorities are interested in the Oxitec mosquitoes and are paying $1.6 million for the program.

Key West is next on the list. Michael Doyle and the Florida Keys Mosquito Control District are very concerned about dengue fever coming back to southern Florida and believe the Oxitec modified mosquitoes are the way to go. Ironically it's the Keys' natural beauty that is responsible for Doyle wanting to use genetically modified mosquitoes, as the Keys' environmental fragility has him concerned about using insecticides. Helicopters are often used to kill mosquito larvae by spraying larvicide on the breeding grounds. But it is much harder to control adult mosquitoes, which must be hit by a droplet of insecticide for them to be killed; this requires large amounts of insecticides and is expensive. Doyle would pay Oxitec between $200,000 and $400,000 for enough *Aedes* eggs to control the dengue-carrying mosquitoes for a year, as opposed to $800,000 for insecticides. "I've looked at all other options for *Aedes aegypti* control, but they're too expensive or environmentally damaging. This sounds like the best option we have going," Doyle concludes. Of course, the Oxitec mosquitoes will solve only his *Aedes* problems, leaving him with 43 other mosquito species to deal with.

Many residents of Key West are concerned about releasing genetically modified mosquitoes and have formed groups to protest the proposed discharge. They have many concerns, such as the fact that the transgenic *Aedes* mosquitoes might have some unforeseen effects on the delicately balanced ecosystems in the Keys or, in the words of Joel Biddle, who once had dengue fever himself, "Corporations like Monsanto and now this Oxitec company are cutting corners and not listening to the scientific community." (21, p. 15)

Hadyn Parry, the CEO of Oxitec, is one of the 35 scientists working for the company and often acts as their spokesperson. He thinks that the public is needlessly concerned about genetic modifications. He says, "Genetic modification is an approach, a tool. It is neither good nor bad. If you take a car and give it to a lunatic it is quite dangerous, but in the right hands it's very useful. You have to see what genetic modification produces and see if the risks and benefits are acceptable. But because there is a public perception issue around genetic modification then we have to overcome the negative perception and that is quite difficult." (21, p. 20)

Doyle would like to start releasing 2 to 6 million Oxitec mosquitoes as soon as possible. He has everything ready but would like governmental approval, which might prove to be the hardest obstacle to overcome. The Food and Drug Administration (FDA) is currently reviewing the application. Doyle could release the mosquitoes without any permission, because there is no law against doing so, but he is not willing to do that. I think Oxitec would also like the credence and legitimacy that the company and its transgenic *Aedes* would receive from the FDA's approval. Luke Alphey says that he welcomes the scrutiny of the regulatory process because he hopes it will assure people of the safety of the Oxitec mosquitoes, particularly because he believes the regulation of transgenic organisms is much stricter than regulation of any other form of insect control. He is also confident that the Oxitec release will be approved by the FDA, explaining, "Our technology has been looked at numerous times by independent regulatory authorities in several different countries and no one has ever found a flaw or a significant risk. I am reasonably confident we will get through that process again." (personal interview)

Anthony James, a geneticist from the University of California, Irvine, is both an Oxitec collaborator and a competitor. He is very concerned that the way the Oxitec mosquitoes have been released may hurt others working in the field, and he fears a public and regulatory backlash that may make it harder for others working in the area to undertake more carefully controlled field studies with extensive information dissemination programs. In a letter published in *Science* magazine, Anthony James acknowledges that there are no industry-wide standards governing the release of genetically modified insects, and he stresses the need for having procedures in place that govern the insects' safe, efficient, ethical, and regulated release. He chastises the Oxitec field releases for jumping the gun and releasing genetically modified mosquitoes before the scientific community has agreed on common standards. James writes, "Thus, although we have not achieved harmonized international standards, as has taken decades for other technologies, we are much closer than most people realize. We recognize the need to ensure that our enthusiasm for the promise of these approaches as powerful public health tools does not outstrip our responsibility to apply scientifically validated and socially acceptable product development practices. The tragedy would be if this important but complex birthing process were to stifle creativity in the development of not only genetics-based solutions, but all truly novel approaches that seek to reduce the serious health threat of diseases such as malaria and dengue fever. We hope that debates over specific circumstances do not cloud the urgent need for the development and deployment of new tools to mitigate these disease scourges." (22, p. 398)

Although James is working with Oxitec, he is not convinced that the company's approach is self-sustaining, and he is looking for an approach that will wipe out 100 percent of the *Aedes* population. He thinks he has found the solution. In a field test in Tapachula, Mexico, which has regular outbreaks of dengue fever, he has been testing his genetically modified *Aedes*. In a variant of the sterile insect technique, his male mosquitoes have been designed to produce female offspring with a genetic defect that makes them flightless. "The flightless females work like a genetic insecticide," says James. So far all his field tests in Mexico have been taken place in large mosquito cages. He is slowly working up to releasing his mosquitoes in the wild. (23)

As long as there are no cures or vaccines for dengue fever, the only way to control the world's fastest-growing infectious disease is to manage the *Aedes* population, either by killing the mosquitoes or by making them inhospitable to the dengue viruses. The current techniques of removing all sources of stagnant water and using limited and targeted insecticide applications are insufficient. It is hard to imagine a solution that doesn't require one of the more controversial mosquito releases. There are some clear advantages to using *Wolbachia* infections or genetically modified self-destructing mosquitoes. They are both species-specific and will not affect any other mosquito species, butterflies, or bees as insecticide applications would, and they can reach places that only male mosquitoes could find. On the other hand, there are no best practices or regulations in place to govern field releases of infected or genetically modified mosquitoes. And because dengue fever occurs in less developed tropical countries, even with the best intentions, conducting field tests and releases of transgenic insects may seem to be taking advantage of them. Finally, it is impossible to predict if there are any long-term consequences, which is a problem because, once released, the mosquitoes cannot be recalled.

REFERENCES

1. Patterson, G. M. (2009). *The mosquito crusades: A history of the American anti-mosquito movement from the Reed Commission to the first Earth Day.* New Brunswick, NJ: Rutgers University Press.
2. Palca, J. (2012). "Eliminate Dengue" team has a deep (lab) bench. *Morning Edition NPR*, June 7.
3. Buchillet, D. (2012). Dengue and dengue-like outbreaks in the past: The case of the Macau epidemic fever of 1874. *Infection Genetics and Evolution 12*, 905–912.
4. Centers for Disease Control and Prevention (2012). Dengue: Symptoms and what to do if you think you have dengue. www.cdc.gov/dengue/symptoms/index.html.
5. Guzman, M. G., Halstead, S. B., Artsob, H., Buchy, P., Jeremy, F., Gubler, D. J., Hunsperger, E., Kroeger, A., Margolis, H. S., Martinez, E., Nathan, M. B., Pelegrino, J. L., Cameron, S., Yoksan, S., and Peeling, R. W. (2010). Dengue: A continuing global threat. *Nature Reviews Microbiology*, S7–S16.
6. Schoggins, J. W., Dorner, M., Feulner, M., Imanaka, N., Murphy, M. Y., Ploss, A., and Rice, C. M. (2012). Dengue reporter viruses reveal viral dynamics in interferon receptor-deficient mice and sensitivity to interferon effectors in vitro. *Proceedings of the National Academy of Sciences of the United States of America 109*, 14610–14615.
7. Sim, S., Ramirez, J. L., and Dimopoulos, G. (2012). Dengue virus infection of the Aedes aegypti salivary gland and chemosensory apparatus induces genes that modulate infection and blood-feeding behavior. *PLOS Pathogens 8*(3), e1002631.
8. World Health Organization (2014). *Fact Sheet #117, Dengue and sever dengue.* http://www.who.int/mediacentre/factsheets/fs117/en/.
9. Sweeney, C. (2012). Buzzkill. *Miami New Times*, May 31, 14–18.
10. Palca, J. (2012). Deconstructing dengue: How old is that mosquito? *Weekend Edition Saturday NPR*, February 11.
11. Adelman, Z. (2012). Personal communication.
12. McInnis, D. (2011). Disease-proofing mosquitoes. IC View. www-angus.ithaca.edu/icview/disease-proofing_mosquitoes-21684/.
13. Adelman, Z. N., Anderson, M. A. E., Morazzani, E. M., and Myles, K. M. (2008). A transgenic sensor strain for monitoring the RNAi pathway in the yellow fever mosquito, Aedes aegypti. *Insect Biochemistry and Molecular Biology 38*, 705–713.
14. Adams, J., Gurney, K. A., and Pendlebury, D. (2012). Neglected tropical diseases. *Thomson Reuters Global Research Report*, June, 1–13.
15. Palca, J. (2012). A scientist's 20-year quest to defeat dengue fever. *Morning Edition NPR*, June 12.
16. Hoffmann, A. A., Montgomery, B. L., Popovici, J., Iturbe-Ormaetxe, I., Johnson, P. H., Muzzi, F., Greenfield, M., Durkan, M., Leong, Y. S., Dong, Y., Cook, H., Axford, J., Callahan, A. G., Kenny, N., Omodei, C., McGraw, E. A., Ryan, P. A., Ritchie, S. A., Turelli, M., and O'Neill, S. L. (2011). Successful establishment of Wolbachia in Aedes populations to suppress dengue transmission. *Nature 476*, 454–457.
17. Walker, T., Johnson, P. H., Moreira, L. A., Iturbe-Ormaetxe, I., Frentiu, F. D., McMeniman, C. J., Leong, Y. S., Dong, Y., Axford, J., Kriesner, P., Lloyd, A. L., Ritchie, S. A., O'Neill, S. L., and Hoffmann, A. A. (2011). The wMel Wolbachia strain blocks dengue and invades caged Aedes aegypti populations. *Nature 476*, 450–453.

18. Hussain, M., Frentiu, F. D., Moreira, L. A., O'Neill, S. L., and Asgari, S. (2011). Wolbachia uses host microRNAs to manipulate host gene expression and facilitate colonization of the dengue vector Aedes aegypti. *Proceedings of the National Academy of Sciences of the United States of America 108*, 9250–9255.
19. Harris, A. F., Nimmo, D., McKemey, A. R., Kelly, N., Scaife, S., Donnelly, C. A., Beech, C., Petrie, W. D., and Alphey, L. (2011). Field performance of engineered male mosquitoes. *Nature Biotechnology 29*, 1034–1037.
20. Oxitec Press Release (2012). Moscamed prepares for the next phase in the development of Oxitec's transgenic mosqitoes in Brazil, http://www.oxitec.com/press-release-moscamed-prepares-next-phase-development-oxitecs-transgenic-mosquitoes-brazil/.
21. Urquhart, C. (2012). Can GM mosquitoes rid the world of a major killer? *Guardian*, July 14, 20.
22. James, A. A. (2011). Genetics-based field studies prioritize safety. *Science 331*, 398.
23. Marsa, L. (2012). Weaponizing mosquitoes to fight tropical diseases. *Pacific Standard*, June 7.

CHAPTER 6

Cancer

Estimated New Cancer Cases and Deaths by Sex, US, 2012

	Male	Female	Total
Estimated new cases	848,170	790,740	1,638,910
Estimated deaths	301,820	275,370	577,190

Source: American Cancer Society

Cancer is an expansionist disease; it invades through tissues, sets up colonies in hostile landscapes, seeking "sanctuary" in one organ and then immigrating to another. It lives desperately, inventively, fiercely, territorially, cannily, and defensively—at times, as if teaching us how to survive. To confront cancer is to encounter a parallel species, one perhaps more adapted to survival than even we are.... A cancer cell is an astonishing perversion of the normal cell. Cancer is a phenomenally successful invader and colonizer in part because it exploits the very features that make us successful as a species or as an organism.

    Siddhartha Mukherjee, *The Emperor of All Maladies: A Biography of Cancer*

Cancer is not a disease. It is a very large group of diseases that can be very different from each other. When normal cells multiply, they follow a highly regulated process in which all the genetic information in the DNA is carefully copied. Growth in normal cells occurs through an ordered sequence of events—the cell cycle, which consists of four distinct phases. Cancer cells are different; they grow without any controls and spread to other parts of the body. The genes that control normal cell growth and division, *oncogenes*, or the genes that inhibit cell division and survival, *tumor suppressor genes*, have been corrupted by mutations.

All cancers divide uncontrollably and can infiltrate and then destroy normal body tissue. This can often have dire consequences. For example, cancer in the brain interferes with the brain's normal functioning and can result in seizures, paralysis, and death. Cancer cells can also upset the chemical balance of their surroundings. Some lung cancers release chemicals that interfere with the control of the calcium concentration and thereby affect nerves and muscles, causing dizziness and weakness.

It is not uncommon that, by the time a cancerous mass is detected, the original cancer cell has been dividing for more than 5 years and that a billion or more cancerous cells are present. As a cancer develops, it changes; the cells mutate and no longer respond to the same drugs and treatment. This makes treating cancer very difficult. Although most cancers form tumors, not all tumors are cancerous. Tumors are abnormal growths of tissue that result from an uncontrolled, progressive multiplication of cells, which serves no physiological function. When the tumor cells do not spread to other parts of the body, the tumor is a benign (not cancerous), and it can often be removed. Cells from malignant tumors use the lymphatic and blood systems to invade nearby tissue and spread to other parts of the body; this is called *metastasis*. (1)

According to Bruce Rothschild, cancer has been around for more than 70 million years. Rothschild, a radiologist at the Northeastern Ohio Universities College of Medicine, should know, for he went looking for and found tumors in dinosaur fossils. He and his team used a portable X-ray machine to scan 10,000 dinosaur vertebrae from more than 700 museum specimens and found bone tumors in close to a hundred duck-billed dinosaurs but in no other species. Rothschild speculates that it was the dinosaurs' conifer diet, which is rich in carcinogenic chemicals, that was responsible for the tumorous growths. (2)

More than 4,600 years ago, the Egyptian physician Imhotep may have been the first to describe cancerous growths. In the translation of the Smith papyrus, which recounts the teachings of Imhotep, we get the description of a bulging mass in the breast that cannot be anything but a breast cancer. (3)

In about 400 B.C., Hippocrates described numerous skin, nose, and breast cancers. One of the names he used for the tumors was *karkinos*, which is Greek for "crab." The Latin word for "crab" is *cancer*; Celsus used the word "cancer" to describe tumors in the first century A.D. Many physicians have written about the origin of the name and the possible connection between crabs and cancer; some saw a similarity between the hardened surface of tumors and the protective shell of crabs, others likened the pain caused by cancers to the bite of a crabs pincer, and yet others thought that the tumorous disease traveled beneath surface of the skin, like a stealthy crab in murky waters. *Onkos*, a Greek word for "mass" or "burden," was occasionally used to describe tumors; this is the origin of the word "oncology," meaning the study of cancer.

In 1761, John Hill published a paper entitled "Cautions against the Immoderate Use of Snuff" in which he linked the tobacco found in snuff to lip, mouth, and throat cancer. The ideas he expressed were correct, although for many reasons the public and his fellow scientists did not accept them. It took almost 200 years before the vast majority of medical doctors saw and understood the connection between lung cancer and tobacco use. But not everyone was convinced. For example, in June 1950, the US surgeon general, Leonard Scheele, commenting on the link between smoking and cancers, said, "The same correlation could be drawn to the intake of milk. . . . Since nothing had been proved there exists no reason why experimental work should be conducted along this line." (3, p. 244)

Fifteen years after Hill described the link between tobacco and throat cancer, Percival Pott, a surgeon at St. Bartholomew's Hospital, London, noted that chimney sweeps had an excessively high rate of scrotal cancer, which he attributed to their exposure to soot. Hill and Pott were among the first to find a cancer-causing agent, a carcinogen. During Pott's lifetime, children as young as 5 were apprenticed as chimney sweeps. In the year that Pott died, 1788, in part in response to his advocacy, the chimney sweepers act was passed; it forbade master sweeps from having apprentices younger than 8. It took until a few decades after Pott's manuscript was published before chimney sweeps became aware of the dangers of soot and learned how to avoid it, especially in their scrotal areas, and before the incidence of scrotal cancer decreased. Unfortunately, the same can't be said for the impact of Hill's work. Centuries after he proved the link between tobacco use and cancer, we are still trying to combat lung cancer by reducing smoking.

The Emperor of All Maladies: A Biography of Cancer by Siddhartha Mukherjee, a former Rhodes scholar, an oncologist, and a professor at Columbia University, attempts to follow the history of cancer, our attitudes toward it, its devious and insidious character (it is a biography, after all), as well as the battles researchers and doctors have fought to control the diseases. (3) In the book, Mukherjee describes how researchers in the United States have spent decades and billions of dollars on the search to find a cure without making much progress. The problem was that government and funding agencies had declared war on cancer. They supported researchers who attacked the disease head-on, and the emphasis was to find new, more radical surgeries, anticancer drugs and radiation treatments. The underlying premise that a systematic all-out search of all known pharmaceuticals and natural products would unearth a cure for one form of cancer and that it would be the root of the ultimate slayer of all cancers was flawed. While effective treatments were found for some cancers like Hodgkin's lymphoma and testicular cancer, from 1950 to 1980 most cancers defied chemotherapy. Fortunately, that has changed, and due to great improvements in molecular biological techniques, including the use of fluorescent proteins, and more focused investigations on individual cancers, the war on cancer has been more successful. Consequently, there has been a significant decrease in cancer mortality since 1995.

Regular screening for many cancers, especially cervical, breast, and colon cancers, has resulted in a significant reduction in fatalities due to these diseases. However, it is impossible to find all cancers in a medical checkup, especially when they are located deep in the body. Cancerous tumors can be diagnosed only by examining excised tumors under a microscope. Computerized tomography and mammograms can be used to indicate the presence of a tumorous growth, but a biopsy is needed so that a microscope can be used to examine the tumor cells themselves to see whether they are cancerous.

Furthermore, there currently is no single technology or exam capable of detecting a cancerous tumor in a patient until the cancer reaches a certain size. This is one of the great difficulties in doing cancer research. How can you observe when a cancerous growth starts, when there are only a limited number of cancerous cells, and how can one follow where the cancerous cells go to when they metastasize? In laboratory model organisms, these are the questions we can answer using fluorescent and bioluminescent proteins.

Fifteen years ago, the majority of medical researchers who were attempting to examine cancerous growths in model organisms such as mice would implant many mice with cancerous cells and then periodically sacrifice them and perform autopsies. Using a microscope, they would search for new tumors in their mice and attempt to quantify the growth and spread of the cancerous cells. This method was far from perfect because many laboratory animals had to be killed, it was very difficult to distinguish between environmental and medicinal effects, and one had to account for the fact that cancerous cells don't grow and react the same in different individual lab animals. However, this was the only way to observe the spatial and temporal spread of cancerous cells in live organisms.

Ideally one would like to monitor the spread of abnormal cells and image their multiplication in living animals, not a series of sacrificed ones, but this is extremely difficult, particularly because most optical methods cannot distinguish between cancerous cells and normal tissue. As a result, they cannot be used to image early-stage tumor growth, or metastasis. Intravital videomicroscopy, that is, the use of video cameras that can act as microscopes inside living organs, has been used, but the procedure is too invasive to lend itself to following tumor growth, progression, and internal metastasis. A much more successful idea has been to label the cancerous cells so that they are the source of light. There are two main approaches to doing this: bioluminescence imaging using firefly luciferase, and fluorescence imaging with green fluorescent protein (GFP)–tagged tumor cells.

The husband-and-wife team of Christopher Contag and Pamela Contag is responsible for developing techniques that image cancer cells with bioluminescence from firefly luciferase. At Stanford University in the early 1990s, Pamela Contag managed to insert the luciferase gene into the bacteria salmonella. There are many different kinds of salmonella, which can cause numerous diseases ranging from typhoid to food poisoning. Pamela Contag was interested in using her transgenic bacteria to see how salmonella affect the host animal cells that they have infected. Christopher Contag, who had read that David Benaron, a pediatrician and engineer, was using laser beams to probe the inner workings of animals, decided to arrange a meeting to determine whether it would be possible to detect the bacterial bioluminescence in live animal cells. The trio hit it off and decided to do a simple trial run in which they inserted a vial of bioluminescing salmonella into a slab of chicken meat, eliminating the need for live animals. The experiment was a success; Benaron's sensitive photon detectors were able to pick up the light emitted by the bacteria from within the chicken. (4)

After the chicken experiment, the researchers infected live rats with transgenic luciferase–producing salmonella. It was extremely difficult to detect the luciferase bioluminescence within the rats, but it did work, and so the Stanford researchers built a highly light-sensitive camera, called a charged couple-device (CCD) camera, to detect and quantify the luminescence given off by the luciferase. Their technique was so successful that in 1997 they formed a biotech company, Xenogen, to commercialize their technology. Pamela Contag left Stanford University to head Xenogen, which was based in Alameda, California; the company produces transgenic mice and rats that have been engineered to emit light when their luciferase-tagged genes are activated, as well as sensitive light recorders and software required for imaging. Xenogen went public in 2004 and was acquired by Caliper Life Sciences in 2006, which in turn was acquired by Perkin Elmer in 2011.

The immune system uses T cells to protect its host from invading pathogens and to eliminate infected or transformed cells. CD8+ cytotoxic T cells are specialized executioners that purge the body of all virally infected cells; as such, they are responsible for eliminating tumors that arise from viral infections, such as human papillomavirus–induced cervical cancer. T cells are less effective at reducing or eliminating tumors that have nonviral origins. According to Brian Rabinovich at the M. D. Anderson Cancer Center in Houston, "'Educating' T cells to home to and kill spontaneous tumors in the same manner as they eliminate virally-infected cells remains the 'holy grail' of tumor immunotherapy." (5, p. 672)

In current immunotherapeutic treatments, between 1 and 40 billion tumor-specific T cells are used. Human clinical trials have shown that a very small number of the T cells used actually infiltrate the cancer and that their therapeutic activity is unpredictable. The injection of large numbers of T cell has numerous drawbacks, including causing respiratory distress. Current efforts at optimizing and understanding immunotherapeutic treatments have been hindered by the lack of imaging techniques that permit the visualization of the small subset of injected T cells that penetrate the tumor. In order to track fewer than a hundred T cells in live mice, Rabinovich and his colleagues have created an enhanced firefly luciferase that made mouse T cells expressing luciferase more than 100 times brighter than those expressing wild-type luciferase. They reported imaging as few as 3 T cells in a tumor; in contrast, the best images obtained with wild-type luciferase showed between 10,000 and 30,000 T cells (figure 6.1). The new enhanced luciferase is now commonly used in labs all around the world and will permit the study of detailed trafficking patterns of the small number of T cells involved in starting an immune response to cancerous tumors. From a therapeutic perspective, it should now be possible to visualize and identify the subsets of T cells constituting less than 0.003 percent of the T cell pool that are responsible for different aspects of antitumor activity and autoimmunity. (6)

Even with the bright mutants, the cells themselves cannot be seen because insufficient photons are produced in a given time to provide an image. The CCD camera therefore counts the photons emitted in a given area over time, and software is used to overlap a pseudo-image upon the mouse.

There are genes that are responsible for the development of new tumors, others that promote the spread of the tumor's cancerous cells to new organs, and then there are some that are required for both. Joan Massagué and his colleagues at the Howard Hughes Medical Institute, Memorial Sloan-Kettering Cancer Center in New York have discovered a set of four

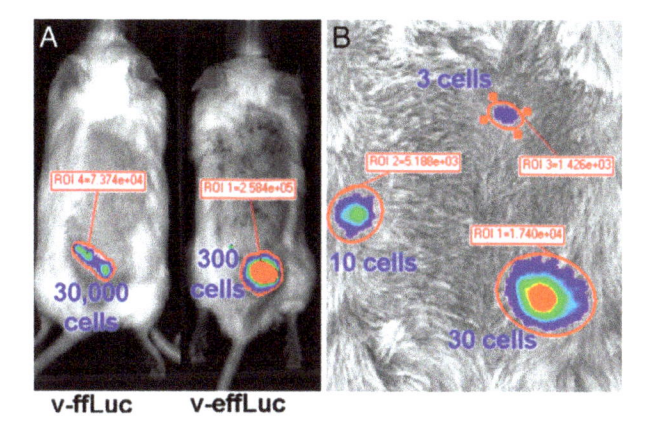

Figure 6.1 It is not possible to image fewer than 10,000 T cells when they have been genetically modified with wild-type firefly luciferase. However, as few as 3 T cells can be detected with Brian Rabinovich's enhanced luciferase. The rat on the far left was injected with 30,000 T cells expressing wild-type luciferase; the enhanced luciferase is responsible for the photon emissions displayed on all the other images. The CCD camera counts the photons emitted by the luciferase and uses a red-to-blue scale to display the emission. (6)

genes that are required for the spread and growth of a breast cancer that metastasizes to the lungs. When the individual genes are turned off, there is a negligible decrease both in the growth of the breast cancer and in its spread to the lungs. However, if all four genes are switched off together, there is a dramatic decrease in both. The four genes act synergistically. By tagging the tumor cells with luciferase, Massagué showed that a combination of all four proteins expressed by these genes increased the amount of cell death, and that they are responsible for trapping metastasizing cells within the capillaries of the lungs (figure 6.2). The amalgamation of these four proteins facilitates the assembly of new tumor blood vessels, the escape of the cancer cells from the breast tumors into circulation, and their subsequent entry into new tissues in the lung. Their ability to catalyze different functions in the primary breast tumor and in the distant lung metastasis distinguishes these four genes from most other oncogenes. Massagué and his coworkers treated mice with a combination of therapies, each known to inhibit one of the four proteins. They observed efficient repression of the primary breast tumors in the mice, as well as a cessation of lung metastasis. As soon as the treatment was stopped, however, the cells trapped within the lung capillaries were released, and metastasis was resumed. (7)

According to Gerhard Christofori from the Institute of Biochemistry and Genetics at the University of Basel, Switzerland, the work marks a technological milestone that will be instrumental in elucidating the combined function of genes during cancer development. In particular, he has

Control   Knockdown

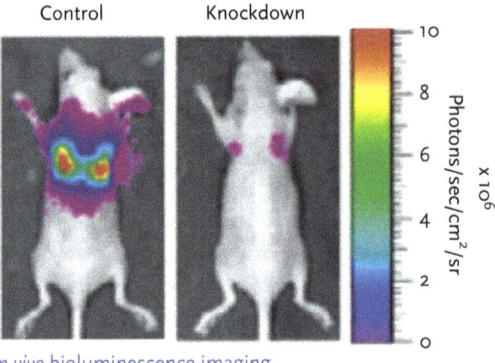

in vivo bioluminescence imaging

Figure 6.2 The mouse on the left is a wild-type control, while the one on the right has four genes switched off; these genes were implicated in the growth of the breast cancer and in its spread to the lungs. Both mice were treated with malignant cells freshly obtained from the pleural fluid of two patients with advanced breast cancer and a diagnosis of lung metastasis. The cells were injected into the tail veins of mice and monitored with serial imaging. Imaging of the luciferase luminescence revealed a significant reduction in lung colonization by the knockdown cells compared with control cells. (7)

asked, are the four genes investigated here "critical players in all subtypes of breast cancer, and possibly in other types of cancer as well? Can the combinatorial treatment, which has proved so successful in the preclinical setting reported by these authors, be further developed for clinical application?" (8, p. 736)

While the Contags were the first to use luciferase to illuminate cancer cells, it was Robert Hoffman and AntiCancer Inc. that first used fluorescent proteins to light up cancer. Hoffman, a biologist with an amazing research career, has published hundreds of papers and is currently both president of AntiCancer Inc. and a member of the University of California San Diego Medical School.

In the early 1990s, Takashi Chishima, who had received his medical degree at Yokohama City University School of Medicine, arrived in San Diego as part of an exchange program between Yokohama and San Diego. Working with Bob Hoffman, he was struggling to find a research area that really interested him. After many months of searching for a project that struck his fancy, Chishima noticed the fluorescent *Caenorhabditis elegans* that Martin Chalfie had created on the cover of *Science*. He was immediately intrigued and knew what his project would be: he was going to create a stable cell line of human cancers that expressed GFP. He worked very hard making hundreds of constructs placed in dishes throughout the laboratory. The project was a success, and Chishima managed to create GFP-expressing cancer cells. When these cells were implanted orthotopically in nude mice, their presence could be detected by their green fluorescence. (*Orthotopically* means simply that the cancer cells were implanted in the "correct organ," for example, lung cancer cells would be implanted in the mouse's lungs, and breast cancer cells in its breast.) Nude mice were chosen as model organisms because they have very weak immune systems and don't reject foreign cells—this means that human cancers could be implanted in nude mice. Furthermore, they have no hair that would interfere with the imaging—with the nude mice, the light is less scattered and attenuated at the skin-air interface. When Chishima and Hoffman and some co-workers published their results in 1997, they reported that they could implant GFP-expressing tumor cells in nude mice and follow the micrometastatic path of the cancer by sacrificing the mouse and looking for fluorescence in its fresh unfixed organs. (9)

The next big breakthrough came a couple of years later, in 1999. It started at a GFP conference in San Diego, which Bob Hoffman attended with Meng Yang, Takashi Chishima's successor from Yokohama. Most large scientific conferences are accompanied by trade shows that exhibit the newest instruments, software, and materials. John S. Fox of Lightools Research was at the GFP conference displaying his equipment as a guest in the Chroma booth. His newest product was the Illumatool, a setup designed for direct viewing or imaging the fluorescence of macro-sized objects, like gels and bacteria in petri dishes.

When Meng Yang saw the Illumatool, with its large stage in which Fox had installed the GFP filter set to observe GFP fluorescence in bacteria, he realized that the device had room for a live mouse. The next morning he and Bob Hoffman sneaked a mouse with a GFP-expressing tumor into the exhibition hall. Lo and behold, using the Illumatool, the fluorescence from a colon tumor was visible in the live mouse. It was a major breakthrough: now tumor growth and metastasis could be followed in a live mouse without having to sacrifice the animal or even anaesthetize it. Figure 6.3 shows a live mouse with two tumors, one expressing green fluorescent protein, and the other red. Fox remembers, "One of the things that struck me was that because of the lack of damaging UV light and brightness, the animals were very comfortable and actually nuzzled the light with their nose. . . . One of the mice had a tumor in the brain, you could see where the skull, tumor and vasculature met. It was a Star Trek moment." (personal interview) The AntiCancer folks were so excited they invited Fox and his Illumatool to join their GFP-expressing mouse at a seminar they were presenting. They did an impromptu show-and-tell, and the mouse and its fluorescent tumor cells quickly became the stars of the conference. According to Hoffman, "That was kind of the end of the meeting, nobody

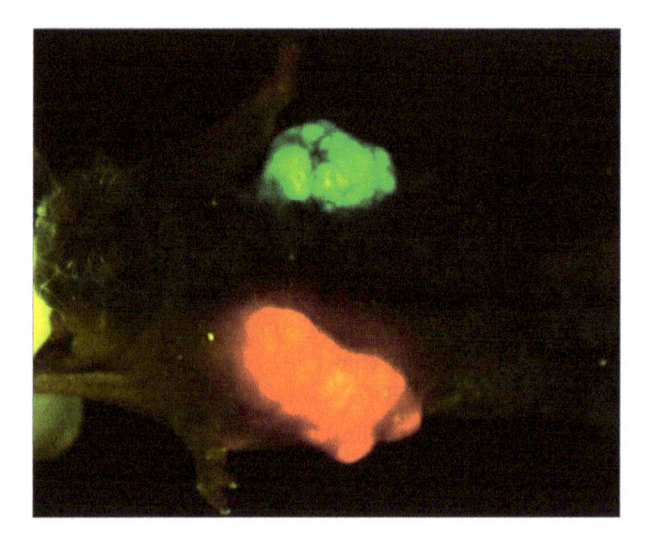

Figure 6.3 Nude mouse with human breast cancer cells expressing green fluorescent protein (top) and red fluorescent protein (bottom). (Image courtesy of AntiCancer Inc.)

wanted to attend lectures anymore, they just wanted to see the mouse." (personal interview)

John Fox and Lightools Research are still in the fluorescence imaging business; if you go to a present-day trade show, you can see Fox with the newest version of the Illumatool, with the LT-99D2-plus dual excitation system that allows you to image two fluorescent colors at the same time, and the Pan-A-See-Ya panoramic imaging system that enables you to see the top view and the two side views of a mouse simultaneously.

Other instruments often used by the current real-time in vivo crowd are confocal fluorescent and high-resolution multiphoton microscopes. Imagine using a microscope to see a tumor inside a mouse; with a conventional microscope, the image of the tumor would be obstructed by the skin and flesh covering the growth, and only the light emitted from the skin would be visible. In confocal microscopes only the signal coming from a predetermined plane is acquired, while the signal coming from the out-of-focus planes is suppressed. Fluorescence images can be obtained using confocal microscopes by illuminating a desired target in the in-focus plane with a laser and using dichroic mirrors to just reflect light of the desired color (e.g., green for GFP-modified tumors) back up to a photodetector. To increase the depth that can be imaged, in multiphoton microscopy short pulses of infrared laser light (recall from the first chapter that infrared light has very little energy but penetrates very efficiently) excite the fluorescent proteins, while the emitted light is detected by photomultipliers.

Following a successful shopping spree at Lightools Research, the researchers at AntiCancer started real-time imaging in live mice. In one of their first sets of experiments, they injected GFP-expressing mouse melanoma cells into the tail veins of 6-week-old nude mice. Subsequent whole-body optical images showed metastatic lesions in the brain, liver, and bone of the mice. This was very exciting because by taking a series of photographs over a period of time, it was now possible to monitor metastases and take real-time, quantitative measurements of tumor growth in each of these organs. The same could be done with fluorescent human colon cancer cells that were surgically implanted into the colons of nude mice. Stimulated by these results, Yang, Hoffman and their colleagues submitted a paper entitled "Whole-Body Optical Imaging of Green Fluorescent Protein–Expressing Tumors and Metastases" to the *Proceedings of the National Academy of Sciences of the United States of America*. The paper contained numerous photographs of metastatic tumors, both human and rodent, in the brain, bone, liver, and lymph nodes of live mice and concluded, "The simple, noninvasive, and highly selective imaging of growing tumors, made possible by strong GFP fluorescence, enables the detailed imaging of tumor growth and metastasis formation. This should facilitate studies of modulators of cancer growth including inhibition by potential chemotherapeutic agents." (*10*, p. 1206) It was a major breakthrough, although the manuscript's publication was nearly derailed because in their introduction the AntiCancer authors favorably compared fluorescent protein technology with luciferase imaging of tumors. This upset some of the luciferase researchers and resulted in their unsuccessfully trying to block the article's publication.

Like many other major scientific breakthroughs, in vivo fluorescence imaging had its detractors, who for many years didn't believe it was possible to observe fluorescence in living organisms; they were convinced that background fluorescence inherent in living cells would obscure the emissions of the fluorescent proteins. Hoffman says, "It got to the point where I would bring mice to meetings with excitation lights and filters, and I would show people. Seeing is believing. I used to have to do that. Of course the new restrictions on handling mice don't allow that anymore." Fortunately, the number of skeptics has dramatically diminished, and the need to transport transgenic mice to conferences in all corners of the world has disappeared.

Cancer cells need oxygen and nutrients to grow beyond a few millimeters in size and spread. They are supplied with these essentials by blood vessels formed from neighboring, healthy tissue. This is called *angiogenesis*. Some chemicals stimulate angiogenesis, and others called angiogenesis inhibitors signal the stop of blood vessel production.

In both animal and clinical studies, angiogenesis inhibitors have successfully stopped new blood vessels from forming, causing some cancers to shrink and die. Avastin was the first antiangiogenesis drug to be approved by the US Food and Drug Administration (FDA) in 2004; since then, the FDA has approved numerous other angiogenic inhibitors for the treatment of cancers such as colorectal and kidney cancer.

It is very difficult to obtain precise data relating angiogenesis to tumor growth and metastasis, especially if mice need to be killed and their organs inspected for each measurement. Researchers at AntiCancer Inc. have been using GFP to study angiogenesis in a way that doesn't require sacrificing animals. They implanted red fluorescent protein–expressing tumors into transgenic GFP mice that express GFP in each and every cell. In this way they have generated lab mice in which all the cells that are derived from the tumors were red fluorescent, and all the cells originating from the host mouse, including blood vessels, gave off green fluorescence (figure 6.4). (11) The technique has been very successful and has been used to light up angiogenesis in bone and soft tissue sarcoma, and in spinal cord cancers that metastasize to the brain. (12) The AntiCancer group has also used a host mouse in which GFP lights up only nascent blood vessels, such as those induced by cancer cells in tumors.

Nude mice are commonly used in disease research. They have a genetic mutation that inhibits their immune systems, which makes them excellent model organisms because they can't muster a rejection response to foreign invaders, such as the implanted red fluorescent cancer cells shown

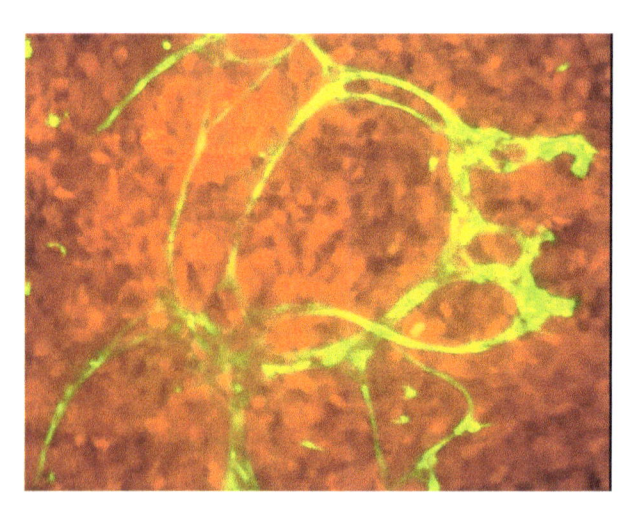

Figure 6.4 Green fluorescent expressing blood vessel growing into a red fluorescent malignant tumor imaged 3 weeks after 10,000 red fluorescent melanoma cells were injected into a GFP mouse. (11) (Image courtesy AntiCancer Inc.)

Figure 6.5 Transgenic GFP nude mouse such as the one used in the angiogenesis experiments shown in Figure 6.4. a. In the top frame the mouse is illuminated with blue light and viewed through a filter that removes the blue light. b. In the bottom frame it has been photographed with ambient lighting. (Image courtesy of AntiCancer Inc.)

in figure 6.3. These hairless mice are the ugliest organisms you can imagine; I doubt that any self-respecting cat would be interested in catching them. But they are so ugly that they are actually quite cute (figure 6.5). Perhaps I am biased, though, because my family has had Shine Shimmer Zimmer and Glowy Glimmer Zimmer, two fluorescent nude mice, as pets. No dogs or cats for the Zimmer family. For years all our friends and guests would get a demonstration of the mice's fluorescence and, of course, a lecture on the uses of fluorescent proteins—I guess I should be grateful that we still have friends. Within a week of getting the mice, my children seem to forget that the mice were fluorescent and had great fun building houses and obstacles for their wrinkly friends. I am not sure that Shine and Glowy ever appreciated that they got to hide in inverted books and Lego castles. Perhaps they wanted something more grandiose, like the 360-degree full-body scanner with a rotating ring of 48 detectors and up to 12 activating lasers that is used to image some of their laboratory mouse brethren. The basic model of this fluorescence-detecting scanner sells for around $300,000 and was awarded the second prize in the 2012 top 10 innovations contest run by *Scientist* magazine. (*13*) It is not your average pet toy.

Cancer begins when mutations of DNA disrupt the cell cycle and uncontrolled cell growth occurs. In February 2008, Atsushi Miyawaki and his group from the RIKEN Basic Science Institute in Japan reported that they had developed a method, known as Fucci (for "fluorescent, ubiquitination-based cell cycle indicator"), that will allow cancer and cell cycle researchers to visualize cell cycle progression. Fucci-modified cells are yellow at the start of replication and then switch to green during S phase and to red during G1. Miyawaki got the idea of naming the technique Fucci (rhymes with Gucci) while waiting for his wife, who was shopping in an upscale shopping mall. The technique has the potential to be very useful in cancer research. The Fucci system allows the real-time visualization of the cell cycle under normal conditions (figure 6.6). (*14*)

In normal cells a series of checks and balances prevents the replication of damaged DNA and the transmission of altered DNA sequences into the next generation. However, by their very nature, cancer cells have compromised cell cycle checkpoints, which means that they continue multiplying despite DNA damage. Furthermore, due to differences in their cell cycle checkpoints, cell death control, and DNA repair, cancer cells vary in their response to anticancer drugs. To further study the impact anticancer drugs have on the cell cycle, Atsushi Miyawaki and his colleagues modified their Fucci system. The new probe, Fucci2, was designed to be used in cultured cells, not in live whole animals; in this way, cancer mechanistic information can be obtained, and many cells can be examined in a very short time. (*15*)

To demonstrate how the Fucci2 cell cycle indicator works, let's look at figure 6.7, which shows the response of a group of cancer cells to varying concentrations of the anticancer drug etoposide. At low concentrations of the drug, the cells cycle as they should and stop at the checkpoint. At intermediate concentrations, the cancer cells override the checkpoint and the nuclei missegregate, which leads to the desired cell death. However, at high concentrations of the anticancer drug, the cells behave differently; they still crash through the checkpoint, but now the DNA is replicated within the nucleus (this is called *endoreplication*). Because these cancer cells now have numerous copies of their genes in the same nucleus, they may be resistant to DNA damage, thereby protecting the cancer cells from the drug. This may be how cancerous cells develop chemotherapeutic resistance.

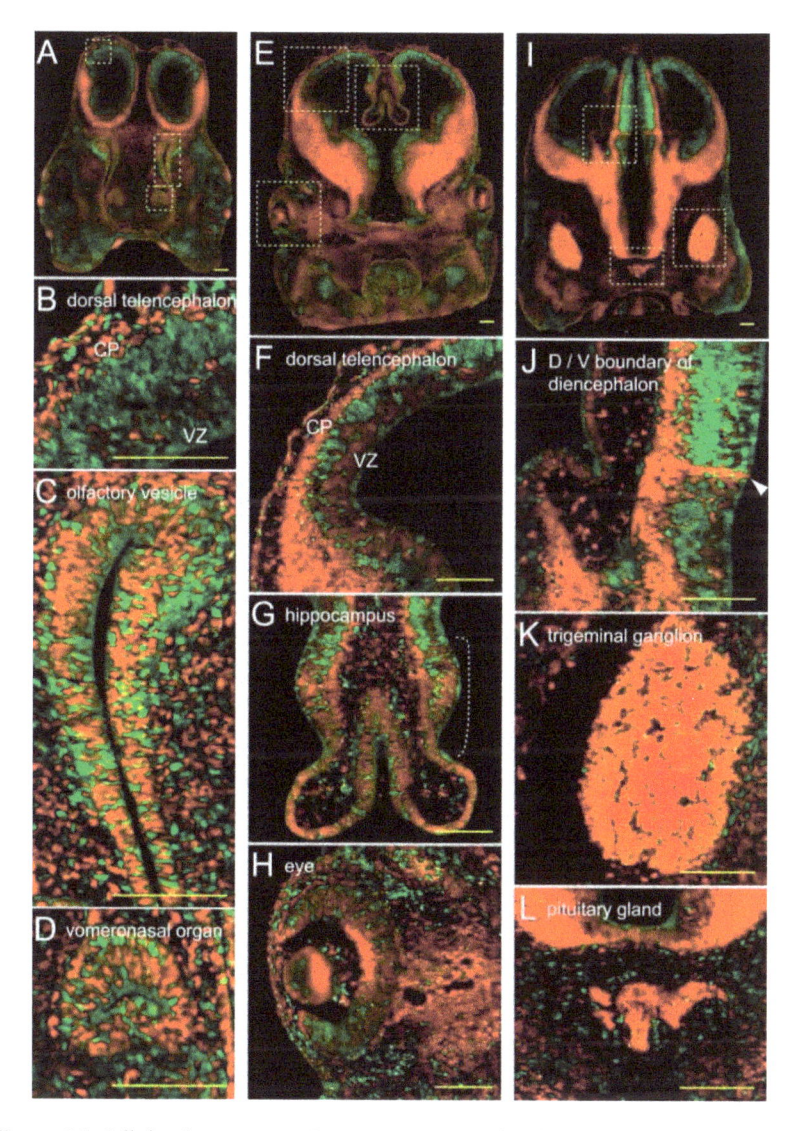

Figure 6.6 Cell development in a Fucci mouse. The colored images shown here demonstrate the equilibrium between cell differentiation and cell proliferation that occurs during the development of the mouse head. Nuclei in the G_1 phase are red, and those in $S/G_2/M$ phases are green. The red and green signals appear to be well balanced at the embryonic stage, but the overall ratio of green-to-red signal decreases as the mice grow. (14)

Close inspection of the cell cycle changes occurring after addition of the chemotherapeutical agent also shows one cancer cell dividing into two daughter cells that react differently to the identical concentrations of the anticancer drug. Once again, it seems that the more you know about cancer, the more complicated it gets. How can you understand what is happening on a macroscopic scale when two such similar cells react so differently

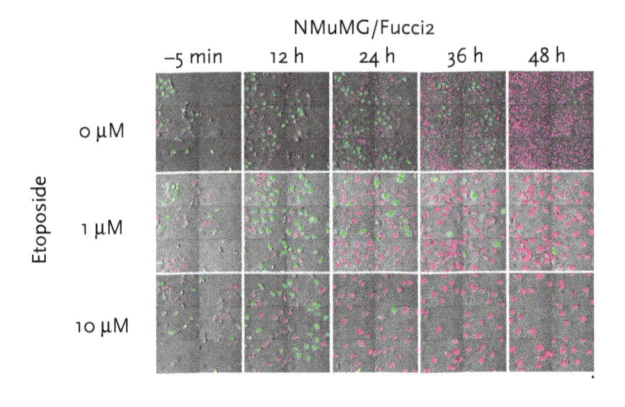

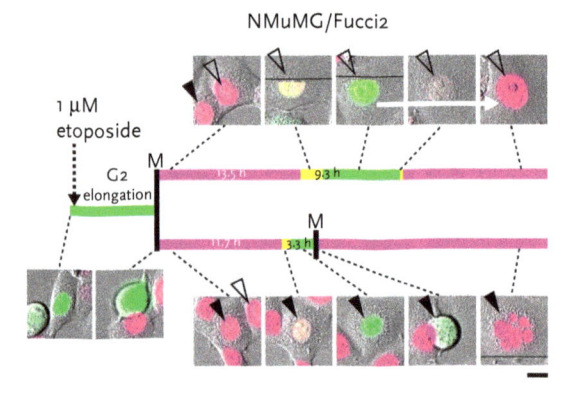

Figure 6.7 Top: Fluorescence images of Fucci2 cancer cells treated with 0, 1, and 10 μM etoposide. At low concentrations of the drug, the cells behave and halt their cell cycle at the checkpoint. At higher concentrations, however, they override the checkpoint, and the nuclei undergo missegregation at intermediate concentrations and endoreplication at high concentrations. Bottom: A Fucci2 cell in the presence of the anticancer drug etoposide was monitored for 48 hours. It divided after a relatively long stay in G2 phase. The two daughter cells experienced very different fates. One cell (labeled with an open arrowhead) changes color but doesn't divide; it underwent genomic replication without cell division (top). The other cell (labeled with a solid arrowhead) underwent nuclear missegregation (bottom). Both cells were killed by the drug, but in very different ways. (15)

to the same drug? Despite the fact that we might never completely understand all the details that link the cell cycle to the initiation and spread of cancerous cells, the use of Fucci2 with large populations of cancer cells monitored as a function of time represents a powerful drug-discovery tool for the development of a new generation of chemotherapies.

A growing number of researchers believe that current cancer treatments such as surgery, chemotherapy, and radiation are inefficient because they do not destroy cancer stem cells. Tumors contain a limited number of cancer stem cells, typically between 1 and 3 percent. These are the cells that drive the tumor's growth; if they aren't destroyed or removed, the cancer will return. The cancer stem cells can be thought of as the roots of the cancer. Similar to weeds in a garden, where pulling them out of the ground may produce good-looking results but be ineffective unless all the roots are removed, tumor regression doesn't necessarily lead to increases in patient survival if the cancer stem cells are not removed.

In August 2012, three papers were published that described using fluorescent proteins to show that a small subset of cells drive tumor growth in brain, (16) gut, (17) and skin (18) cancers. If these findings apply to all other cancers, Luis Parada from the University of Texas Southwestern Medical Center in Dallas thinks that "there is going to be a paradigm shift in the way chemotherapy efficiency is evaluated and how therapeutics are developed." (19, p. 13) Parada should know because it is his group that proved the existence of a subset of cells, cancer stem cells, that is responsible for glioblastoma multiforme, the most common primary malignant brain tumor. Patients with glioblastoma multiforme have a median post-treatment survival of about a year, which is in large part due to its recurrence after surgical removal. In his studies Parada used a technique that labels a subset of quiescent glial cells that exhibit stem cell–like behavior with GFP. His team found that after the growth of the brain tumors in mice is arrested with the drug temozolomide, new tumor growth occurs, originating from the GFP-labeled glial cells. Perhaps more significantly, if these GFP-expressing cells are removed, tumor growth is significantly impaired (figure 6.8). These GFP-expressing glial cells are cancer stem cells that lurk quietly in the tumor, unaffected by the tumor-destroying temozolomide, and are ready to spit out short-lived populations of highly proliferative tumor cells. (16)

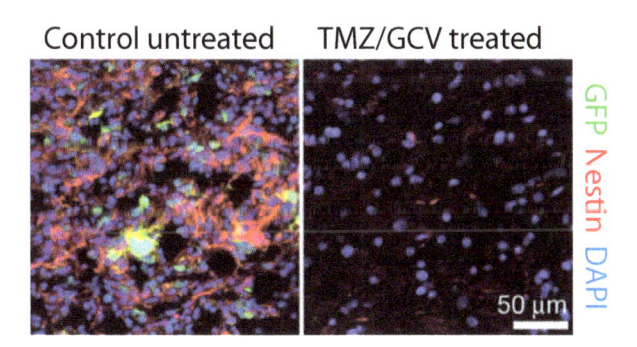

Figure 6.8 Untreated mouse brains have green fluorescent glioma cells (left), which act as cancer stem cells by sustaining long-term tumor growth through the production of transient populations of highly proliferative tumor cells. After the fluorescent stem cells have been removed by ganciclovir treatment (right), tumor development is significantly impeded. (16)

Acknowledging the existence of cancer stem cells and understanding their cancer-initiating properties maybe crucial in designing effective anti-cancer treatments, particularly because recent evidence has been found that suggests that cancer stem cells are resistant to molecular-targeting anticancer drugs. (20) The drug Gleevec may be an excellent example of a molecular-targeting therapy that is foiled by cancer stem cells. It is a very effective drug for chronic myelogenous leukemia, and patients treated with it go into remission. However, the leukemia stem cells are resistant to Gleevec, and the cancer returns after the drug is no longer taken. In transgenic GFP mice, supplementation of Gleevec treatment with agents that target leukemia stem cells has been very successful and has been suggested as a new way of improving current Gleevex treatments in humans. (21)

Herceptin, which has a wholesale price of $54,000 for a year's treatment, is a hugely successful breast cancer drug designed to interfere with the HER2 receptor. A surprising report published in 2008 by researchers from the University of Pittsburgh showed that Herceptin also appeared to benefit patients who are missing the HER2 receptor. (22) This finding has subsequently been explained by the observation that some breast tumors express the HER2 receptors only in their cancer stem cells. They are HER2-negative so the Herceptin does not attack the non-stem cancer cells, but, perhaps more important, it attacks the HER2 receptors on the cancer stem cells. (23)

The two most important properties of stem cells are their ability to undergo self-renewal and to differentiate into new cell types. In comparison to most other stem cells, cancer stem cells rarely differentiate. Mickie Bhatia, from the McMaster Stem Cell and Cancer Research Institute in Hamilton, Ontario, Canada, has developed a GFP-based screen to find drugs that force cancer stem cells to differentiate, thereby exhausting the supply of self-renewing cells. "Now we can test thousands of compounds, eventually defining a candidate drug that has little effect on normal stem cells but kills the cells that start the tumor," said Bhatia. (24) The most promising compound identified so far has been thioridazine, a discontinued antipsychotic drug. In a neat bit of detective work, Bhatia and colleagues were able to work out why thioridazine targets cancer stem cells. They discovered that thioridazine acts on dopamine receptors in schizophrenics and that dopaminergic-deficient Parkinson's patients have lower incidences of cancer, as do patients taking other medications that target dopamine receptors. This led them to hypothesize and then confirm that dopamine receptors were to be found on the surface of the rare tumor-initiating stem cells but not on the blood stem cells, which is why the cancer stem cells were the only cells affected by thioridazine. (25)

Many carcinomas will reappear in new organs years after surgery. Most current experimental evidence indicates that tumor cells disseminate from early lesions and then undergo an extended period of dormancy in their target organs before being activated as active tumor cells. Filippo Giancotti and a team of researchers from the Sloan-Kettering Institute for Cancer Research have been interested in finding the protein responsible for inducing dormant breast cancer cells to undergo the change to active lung cancer stem cells. They found that only cancer cells that carry the gene for the protein Coco become fully proliferative in the lung. Coco is the on switch; it blocks regulators that prevent cancer cells from transitioning to proliferative stem cancer cells. Without Coco, the regulatory proteins can do their job, but as soon as Coco is expressed, the regulatory proteins can no longer suppress the cancer stem cell activation. Figure 6.9 shows some dormant GFP-labeled breast cancer cells lurking in the lung tissue, waiting for Coco to initiate their transition to fully metastatic lung cancer cells. (26)

Miriam Fein and Mikala Egeblad from the Cold Spring Harbor Laboratories in New York have written a review entitled "Caught in the Act: Revealing the Metastatic Process by Live Imaging," (27) and although its focus is on metastasis, it shows how much real-time imaging has changed medical research as a whole. Traditionally, metastasic studies could only rely on measurements of newly formed metastatic growths; researchers interested in metastasis could only dream of interrogating a cancerous cell as it exited the primary tumor, entered the blood or lymphatic system, got swept along to the secondary site where it left the vessels, invaded the surrounding tissue, and started a secondary tumor

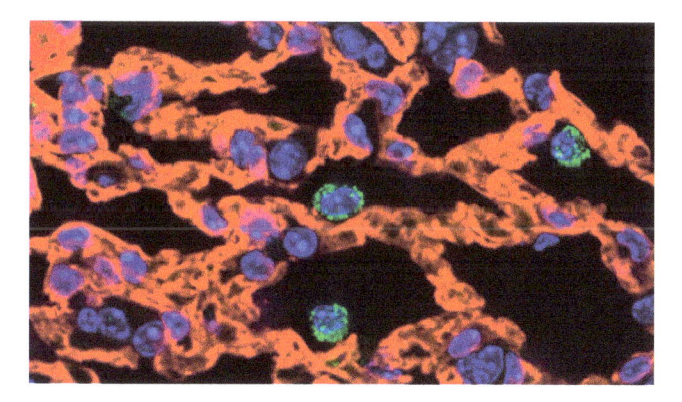

Figure 6.9 Expression of the protein Coco activates GFP-labeled breast cancer cells that have been lying dormant in lung tissue. It is a newly discovered on switch that may be responsible for breast cancers that reappear in the lungs many years after surgery. (26)

growth. That has all changed. Now we can see how cancer cells migrate as single cells, are drawn to chemoattractants, and react to their local environments. Figure 6.10 summarizes how fluorescent proteins have helped us illuminate the dynamics of nearly every step of the metastatic process.

All the methods I have described in this chapter are fluorescent protein–based techniques that were designed to monitor the growth and spread of cancerous cells. They have become very important tools in cancer research and in drug discovery. They have not yet made their way into the clinic, but that is going to change. Before long, we are likely to see fluorescent viruses assisting surgeons in the complete removal of cancerous growths and cancer-targeting fluorescent bacteria supplementing current chemotherapeutic methods.

Both the GFP-based and luciferase-based methods for following the movement and spread of cancer cells rely on genetically modifying cancer cells. This is useful in studying the behavior of cancer cells in living model organisms, but it is not very practical in the direct treatment of cancer in humans. However, back in 2003, Aladar Szalay, who holds positions at both the University of Würzburg in Germany and the School of Medicine at Loma Linda University in California, planted the seeds for a new method for monitoring primary tumors and metastases in animals that might soon be used in humans. The technique is based on the observation that some microorganisms, such as viruses and bacteria, are present in tumor tissue excised from human patients but are absent in the rest of the body. Szalay and his colleagues have genetically modified these microorganisms with light-emitting proteins such as firefly luciferase and GFP, so that they can visualize their movement from the bloodstream into the tumor region and their replication in solid tumors. (28)

Szalay and his colleagues discovered that light-emitting bacteria colonized tumors to such an extent that, a week after injection, none of the bacteria were found in the blood or internal organs. They were all localized at the tumor. In fact, they were so restricted to the tumor that they were not released into the circulation at all, or at least in insufficient numbers to be able to colonize other newly implanted tumors.

Szalay's work has shown that while most of the GFP-labeled bacteria collect in the center of the tumors, GFP-labeled viruses aggregate at the periphery of the tumors, where the fast-dividing cells are located. This led him to speculate that GFP fluorescence from labeled viruses may be used as a marker for identifying tumor margins to facilitate precise tumor removal during surgery. He was correct, and currently numerous groups are using fluorescence-guided surgical techniques in mice.

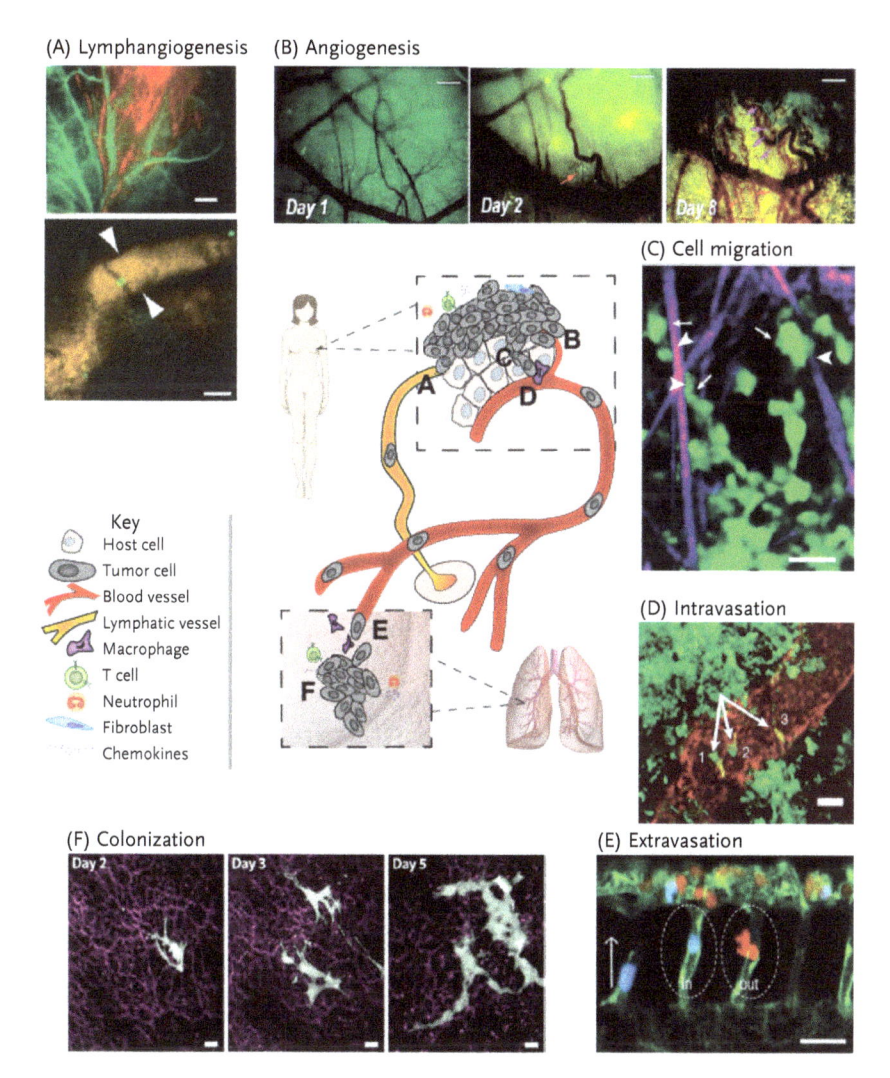

Figure 6.10 The central drawing summarizes the metastatic processes imaged with fluorescent proteins and shown in A through F. (A) Tumor blood (green) and lymphatic (red) vessels are dilated in comparison to normal vessels found in the ear (top). Green fluorescent fibrosarcoma traveling in a lymphatic vessel from the primary tumor to the cervical lymph node (bottom). (B) Blood vessels growing into mammary carcinoma cells implanted in a mouse model. (C) GFP-labeled breast cancer cells move along collagen fibers (purple). Arrows point to carcinoma cells, and arrowheads point to cell-matrix interactions. (D) GFP-labeled breast cancer cells escaping primary tumor and entering into a dextran-labeled blood vessel (red) imaged by multiphoton microscopy. Three cells that have crossed into the blood vessel are shown in yellow and indicated by arrows (scale bar: 25 μm). (E) Confocal image of human breast cancer cells in zebrafish embryos. Some of the cells have been genetically modified to be particularly efficient at breaking out of the blood vessels. They express RFP and are colored red, while the nonextravasating wild types express cyan fluorescent protein and are colored blue (scale bar: 200 μm). (F) Dendra, another fluorescent protein, expressing colorectal carcinoma cells colonizing a mouse liver. Fibrillar collagens were colored purple (scale bar: 20 μm). (27)

To confirm that the viruses really do find all the tumors, Bob Hoffman and his AntiCancer team injected red fluorescent human colon cancer cells into the peritoneal cavity of a mouse. Once the resultant red fluorescent tumors had spread in the mouse, the researchers injected the mouse with a virus strain that collects in malignant tissue and expresses GFP only in tumor tissues. Five days after viral administration, all red fluorescent tumors were labeled by the green fluorescent viruses (figure 6.11). (29) Having proved that their viruses label all colon tumors and that there are no false positives, the AntiCancer researchers have gone on to show that fluorescence-guided surgery improves recovery rates and removal of orthotopically placed human colon cancers in mouse models. (30)

Mark Tangney from the Cancer Research Center at the University of College Cork in Ireland has other ambitions for his tumor-seeking bacteria (figure 6.12). (31) According to Michelle Cronin, a research fellow working with Tangney, they want to use their bioluminescent tumor-colonizing bacteria to deliver chemotherapy directly to the cancer cells. "We can now genetically engineer these bacteria so that they will pump out anti-cancer agents specifically inside tumors," she says. "And we are testing our engineered strains in numerous tumor models and as we

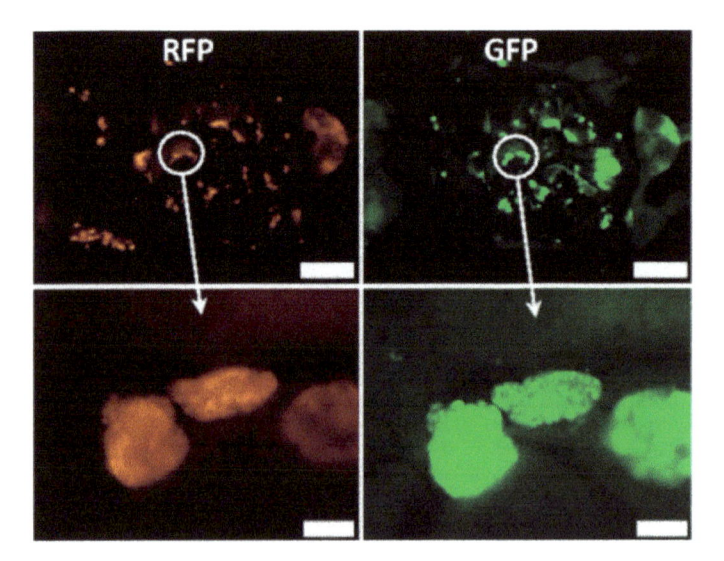

Figure 6.11 Twelve days after red fluorescent human colorectal cancer cells were injected into a mouse, tumor-targeting viruses that fluoresce only in the presence of malignant cells were added. Using a filter to image the red fluorescence, all the cancerous cells were located (left images). The green channel showed that all (red fluorescent) tumors were engulfed by GFP expressing viruses (right images). These images confirm that the GFP-expressing viruses can be used to find and delineate the surfaces of malignant tumors during surgery. (29) (Scale bars: upper, 10 mm; lower, 500 μm.)

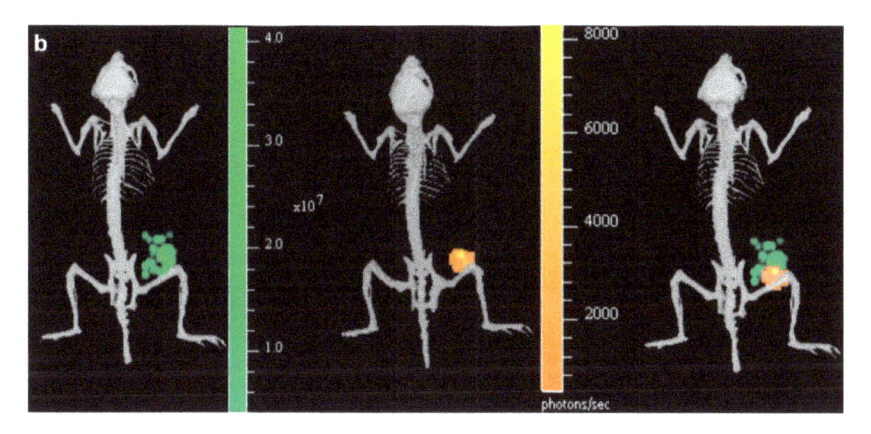

Figure 6.12 Three-dimensional overlay of bacterial (colored orange and arising from bacterial luciferase) and tumor (colored green and arising from firefly luciferase) bioluminescence. The overlap confirms that the bacteria collect at the tumor and that they could act as chemotherapy delivery agents. (*31*)

function as a translational research center, we hope to progress the concept to pre-clinical trials." (*32*)

The bacteria don't always have to act as tumor-targeting drug delivery vehicles; sometimes they can even attack the tumors themselves. There is some well-aged precedent for this. In 1893, using only bacteria, William Coley successfully treated John Ficken, a 16-year-old boy with a massive abdominal tumor. Every few days Coley injected a concoction of *Streptococcus pyogenes* into the tumor, inducing an infection that shrunk the growth until the tumor disappeared. Many types of bacteria, like *S. pyogenes*, target malignant tumors, but most of these bacteria live in oxygen-free environments, and so they only grow in the necrotic parts of the tumor. This severely limits the utility of bacterial cancer therapy, and it never caught on. But the partial successes weren't forgotten, and more than a hundred years later, the AntiCancer group revisited bacterial cancer therapy.

When I asked Hoffman to list his favorite experiments involving GFP, he had quite a list. Most of them involved real-time imaging, and one was responsible for AntiCancer's interest in reviving Coley's 1893 experiments. Hoffman recalled, "It was unforgettable to watch GFP-labeled bacteria traverse the intestinal tract noninvasively; we knew where they were every step of the way, going in from the mouth and out toward the other end. It was this experiment that led us to develop tumor-targeting bacteria and label them with GFP so we could follow them as they targeted the tumor." (personal interview)

To improve on Coley's bacteria that targeted only oxygen-deprived tumor cells, AntiCancer Inc. has developed a strain of *Salmonella*

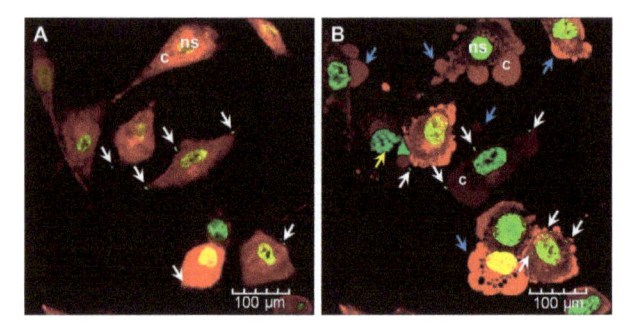

Figure 6.13 In vitro imaging of dual-color lung cancer cells expressing GFP in the nucleus (ns) and DsRed in the cytoplasm (c) during bacterial treatment. The image on the left shows the cancer cells 10 minutes after treatment with bacteria; white arrows show bacteria that have invaded the cancer cells. The image on the right shows the same cancer cells 45 minutes after treatment with *S. typhimurium*. The cancer cells have expanded, burst, and lost viability. The yellow arrow shows a damaged nucleus. (*33*) (Image courtesy of AntiCancer Inc.)

typhimurium, which grows in both the presence and the absence of oxygen. This strain of *S. typhimurium* targets both the viable and the necrotic areas of the tumors and inhibits tumor growth in mice.

Vion Pharmaceuticals has developed a related *Salmonella* strain that can safely be administered to patients. In clinical studies, tumor colonization was observed with this strain, but it did not exhibit the desired antitumor effects in humans.

AntiCancer's strain has been designed to be more tumor virulent than the Vion strain, but it has yet to be tested in a clinical trial. The AntiCancer researchers and their collaborators are optimistic and sought to further understand the mechanism of the bacterial cancer-cell killing in mice. Perhaps knowledge of the mechanism would allow them to optimize it for clinical applications. To visualize the cancer-cell killing, GFP-expressing *S. typhimurium* were injected into the tail veins of mice containing lung cancer cells that expressed GFP in the nucleus and DsRed in the cytoplasm. Within minutes of the tail vein injection, the bacteria targeted the tumor and killed the cancer cells by causing them to expand and burst; similar behavior was observed in in vitro experiments (figure 6.13). (*33*)

The rapid increase in the use of fluorescent proteins in cancer research and the growing importance of real-time imaging in medicine lead us to some interesting questions about the future. Twenty years from now, will fluorescent proteins be as ubiquitous in hospitals as they are in research labs? Will fluorescent viruses be used to define the borders of cancerous tumors in laparoscopic surgery? Will fluorescent bacteria hunt and kill malignant cells in cancer patients, or will microscopic cameras patrol our blood and lymphatic systems looking for fluorescent responses to healthy cells going bad? I have to say, I think the future for fluorescent proteins in cancer treatment is quite bright.

REFERENCES

1. Zimmer, M. (2005). *Glowing genes: A revolution in biotechnology.* Amherst, NY: Prometheus Books.
2. Rothschild, B. M., Tanke, D. H., Helbling, M., and Martin, L. D. (2003). Epidemiologic study of tumors in dinosaurs. *Naturwissenschaften 90*, 495–500.
3. Mukherjee, S. (2010). *The emperor of all maladies: A biography of cancer.* Large print ed. Waterville, ME: Thorndike Press.
4. Kruszelnicki, K. S. (2002). Light of life. www.abc.net.au/science/k2/moments/s587114.htm.
5. Rabinovich, B.A., and Radu, G. (2010). Imaging adoptive cell transfer based cancer immunotherapy. *Current Pharmaceutical Biotechnology 11*, 672–684.
6. Rabinovich, B. A., Ye, Y., Etto, T., Chen, J. Q., Levitsky, H. I., Overwijk, W. W., Cooper, L. J. N., Gelovani, J., and Hwu, P. (2008). Visualizing fewer than 10 mouse T cells with an enhanced firefly luciferase in immunocompetent mouse models of cancer. *Proceedings of the National Academy of Sciences of the United States 105*, 14342–14346.
7. Gupta, G. P., Nguyen, D. X., Chiang, A. C., Bos, P. D., Kim, J. Y., Nadal, C., Gomis, R. R., Manova-Todorova, K., and Massagué, J. (2007). Mediators of vascular remodelling co-opted for sequential steps in lung metastasis. *Nature 446*, 765–770.
8. Christofori, G. (2007). Cancer: Division of labour. *Nature 446*, 735–736.
9. Chishima, T., Miyagi, Y., Wang, X., Yamaoka, H., Shimada, H., Moossa, A. R., and Hoffman, R. M. (1997). Cancer invasion and micrometastasis visualized in live tissue by green fluorescent protein expression. *Cancer Research 57*, 2042–2047.
10. Yang, M., Baranov, E., Jiang, P., Sun, F. X., Li, X. M., Li, L. N., Hasegawa, S., Bouvet, M., Al-Tuwaijri, M., Chishima, T., Shimada, H., Moossa, A. R., Penman, S., and Hoffman, R. M. (2000). Whole-body optical imaging of green fluorescent protein-expressing tumors and metastases. *Proceedings of the National Academy of Sciences of the United States of America 97*, 1206–1211.
11. Yang, M., Li, L. N., Jiang, P., Moossa, A. R., Penman, S., and Hoffman, R. M. (2003). Dual-color fluorescence imaging distinguishes tumor cells from induced host angiogenic vessels and stromal cells. *Proceedings of the National Academy of Sciences of the United States of America 100*, 14259–14262.
12. Hayashi, K., Yamauchi, K., Yamamoto, N., Tsuchiya, H., Tomita, K., Bouvet, M., Wessels, J., and Hoffman, R. M. (2009). A color-coded orthotopic nude mouse treatment model of brain-metastatic paralyzing spinal cord cancer that induces angiogenesis and neurogenesis. *Cell Proliferation 42*, 75–82.
13. Staff, The Scientist (2012). Top ten innovations. *The Scientist 1*, 43.
14. Sakaue-Sawano, A., Kurokawa, H., Morimura, T., Hanyu, A., Hama, H., Osawa, H., Kashiwagi, S., Fukami, K., Miyata, T., Miyoshi, H., Imamura, T., Ogawa, M., Masai, H., and Miyawaki, A. (2008). Visualizing spatiotemporal dynamics of multicellular cell-cycle progression. *Cell 132*, 487–498.

15. Sakaue-Sawano, A., Kobayashi, T., Ohtawa, K., and Miyawaki, A. (2011). Drug-induced cell cycle modulation leading to cell-cycle arrest, nuclear mis-segregation, or endoreplication. *BMC Cell Biology 12*, 2–15.
16. Chen, J., Li, Y., Yu, T.-S., McKay, R. M., Burns, D. K., Kernie, S. G., and Parada, L. F. (2012). A restricted cell population propagates glioblastoma growth after chemotherapy. *Nature 488*, 522–526.
17. Schepers, A. G., Snippert, H. J., Stange, D. E., van den Born, M., van Es, J. H., van de Wetering, M., and Clevers, H. (2012). Lineage tracing reveals Lgr5+ stem cell activity in mouse intestinal adenomas. *Science 337*, 730–735.
18. Driessens, G., Beck, B., Caauwe, A., Simons, B. D., and Blanpain, C. (2012). Defining the mode of tumour growth by clonal analysis. *Nature 488*, 527–530.
19. Baker, M. (2012). Cancer stem cell tracked, *Nature 488*, 13–14.
20. Liu, S. L., Korkaya, H., and Wicha, M. S. (2012). Are cancer stem cells ready for prime time? *The Scientist 26*, 32–37.
21. Li, L., Wang, L. S., Li, L., Wang, Z. Q., Ho, Y. W., McDonald, T., Holyoake, T. L., Chen, W. Y., and Bhatia, R. (2012). Activation of p53 by SIRT1 inhibition enhances elimination of CML leukemia stem cells in combination with imatinib. *Cancer Cell 21*, 266–281.
22. Paik, S., Kim, C., and Wolmark, N. (2008). HER2 status and benefit from adjuvant trastuzumab in breast cancer. *New England Journal of Medicine 358*, 1409–1411.
23. Liu, S. L., and Wicha, M. S. (2010). Targeting breast cancer stem cells. *Journal of Clinical Oncology 28*, 4006–4012.
24. Kankudti, A. (2012). Cancer stem cells targeted by anti-psychotic drug. *Medical Daily*. www.medicaldaily.com/news/20120526/10051/cancer-thioridazine-stem-cells.htm.
25. Sachlos, E., Risueno, R. M., Laronde, S., Shapovalova, Z., Lee, J. H., Russell, J., Malig, M., McNicol, J. D., Fiebig-Comyn, A., Graham, M., Levadoux-Martin, M., Lee, J. B., Giacomelli, A. O., Hassell, J. A., Fischer-Russell, D., Trus, M. R., Foley, R., Leber, B., Xenocostas, A., Brown, E. D., Collins, T. J., and Bhatia, M. (2012). Identification of drugs including a dopamine receptor antagonist that selectively target cancer stem cells. *Cell 149*, 1284–1297.
26. Gao, H., Chakraborty, G., Lee-Lim, A. P., Mo, Q. X., Decker, M., Vonica, A., Shen, R. L., Brogi, E., Brivanlou, A. H., and Giancotti, F. G. (2012). The BMP inhibitor Coco reactivates breast cancer cells at lung metastatic sites. *Cell 150*, 764–779.
27. Fein, M. R., and Egeblad, M. (2013). Caught in the act: Revealing the metastatic process by live imaging. *Disease Models & Mechanisms 6*, 580–593.
28. Yu, Y. A., Timiryasova, T., Zhang, Q., Beltz, R., and Szalay, A. A. (2003). Optical imaging: Bacteria, viruses, and mammalian cells encoding light-emitting proteins reveal the locations of primary tumors and metastases in animals. *Analytical and Bioanalytical Chemistry 377*, 964–972.
29. Kishimoto, H., Zhao, M., Hayashi, K., Urata, Y., Tanaka, N., Fujiwara, T., Penman, S., and Hoffman, R. M. (2009). In vivo internal tumor illumination by telomerase-dependent adenoviral GFP for precise surgical navigation. *Proceedings of the National Academy of Sciences of the United States 106*, 14514–14517.
30. Metildi, C. A., Kaushal, S., Hardamon, C. R., Snyder, C. S., Pu, M. Y., Messer, K. S., Talamini, M. A., Hoffman, R. M., and Bouvet, M. (2012). Fluorescence-guided surgery allows for more complete resection of pancreatic cancer, resulting in longer disease-free survival compared with standard surgery in orthotopic mouse, models. *Journal of the American College of Surgeons 215*, 126–135.

31. Cronin, M., Akin, A. R., Collins, S. A., Meganck, J., Kim, J.-B., Baban, C. K., Joyce, S. A., van Dam, G. M., Zhang, N., van Sinderen, D., O'Sullivan, G. C., Kasahara, N., Gahan, C. G., Francis, K. P., and Tangney, M. (2012). High resolution *in vivo* bioluminescent imaging for the study of bacterial tumour targeting. *PLOS One* 7, e30940.
32. Pittman, D. (2012) Engineered Bacteria Target Tumors, *Biotechniques*, http://www.biotechniques.com/news/biotechniquesNews/biotechniques-326441.html.
33. Uchugonova, A., Zhao, M., Zhang, Y., Weinigel, M., Konig, K., and Hoffman, R. M. (2012). Cancer-cell killing by engineered Salmonella imaged by multiphoton tomography in live mice. *Anticancer Research 32*, 4331–4337.

CHAPTER 7

Influenza

Outbreak	Number infected	Deaths	Case fatality rate
Seasonal flu	~100 million+	~50,000	<0.1%
2009 (swine)	~60 million	~15,000	<0.1%
2006 (avian)	115	79	~60%
1918	~500 million	~50 million	2.5–10%

> Some coughed so hard that autopsies would later show they had torn apart abdominal muscles and rib cartilage. And many of the men writhed in agony or delirium; nearly all those able to communicate complained of headache, as if someone were hammering a wedge into their skulls just behind the eyes, and body aches so intense they felt like bones breaking. A few were vomiting. Finally the skin of some of the sailors had turned unusual colors.
>
> <div align="right">John M. Barry, <i>The Great Influenza</i></div>

As winter approaches, the cold forces us into the warmth of our homes, where interactions with our fellow humans are magnified by increased close contact. These conditions are ideal for the spread of the influenza (flu) virus, which is transmitted mainly by sneezes. A tear-inducing, room-resonating sneeze is a viral ejection of massive proportions. It starts with a tickling feeling in the nose, grows with each vain attempt at repression of the inevitable, and ends with the violent expulsion of as many as 40,000 projectiles travelling up to 200 miles an hour. Each droplet expelled into the room can contain as many as 100 million flu viruses. The virus-laden water bombs are also released in much less dramatic fashion when someone with the flu talks or coughs. After expulsion from the infected body, the larger droplets fall to ground, while the smaller ones can remain airborne for days. Most flu viruses are spread when these droplets directly enter someone's mouth or nose. It is a common misconception that flu infections come from touching doorknobs, handrails, and so forth; actually, infection from droplets that have landed on a surface is fairly rare. During the summer, we spend more time outdoors, where we are less likely to breathe in virus-laden droplets. Consequently, outbreaks of the flu are generally limited to the winter season. The outbreaks peak in about 3 weeks and take another 3 weeks to dissipate. The symptoms of these seasonal flu outbreaks are similar to those of a cold but are more severe. They can include fever, cough, sore throat, aching muscles and joints, headache, and general malaise, the severity of which depends on the influenza viruses involved. In the period between 1976 and 2006, annual flu-associated deaths in the United States ranged from a low of 3,000 to a high of 50,000 people. Older people, young children, pregnant woman, and people with asthma are particularly susceptible to flu viruses and are at greater risk for serious complications. (1)

There are currently two types of FDA-approved drugs for influenza infections, one that prevents the unwrapping of the virus and thwarts its entry into cells, and another that blocks viral spreading. Many research groups all over the world have been struggling to create fluorescent influenza viruses so that they can study them and observe the effects of the anti-flu drugs on their behavior. Finally, in July 2010, Adolfo Garcia-Sastre, a microbiologist at the Global Health and Emerging Pathogens Institute at Mount Sinai School of Medicine, New York, reported a functional GFP-expressing influenza virus. It was the first time a flu virus was created that could replicate, infect mice, and make GFP. Using the GFP tagged influenza virus, Garcia-Sastre and his coworkers were able to follow the dynamics of influenza virus infection and the cell types that were infected in live mice. They used their fluorescent flu viruses to examine how antiviral drugs affect the immune response to the influenza virus. The GFP-labeled viruses showed that the immune response of some cell types was changed by the anti-flu medicines. (2) Since the report by Garcia-Sastre and his colleagues was published, at least 30 other papers have been published using the fluorescent flu viruses, most of them looking for new, improved antiflu drugs.

There is a definitive need for new techniques to monitor the actions of the flu virus and to find new anti-influenza drugs. Noelle-Angelique Molinari, a statistician at the Centers for Disease Control and Prevention in Atlanta, Georgia, and her co-workers have analyzed the annual burden and cost of seasonal influenza for the United States. They found that every year an average of 610,660 years of life are lost, 3.1 million hospitalized days are incurred, and 31.4 million outpatient visits take place due to flu outbreaks. The direct medical costs average $10.4 billion annually, and the projected lost earnings due to illness and loss of life amount to $16.3 billion annually. (3, p. 5086) These numbers are staggering, but they are insignificant compared with the human and economic consequences of a flu pandemic.

At least four times in recorded history, a flu outbreak has swept across the world, infecting a significant portion of its population. Such outbreaks are known as *pandemics*. The most recent flu pandemic occurred in 2009 and was known as the swine flu pandemic because the virus had been circulating among pigs before infecting humans. However, it is more accurately a bird flu because it originated in birds. In fact, all human flu viruses ultimately originate from bird populations, which makes naming them hard. Clearly, a more accurate naming procedure is required.

There are three types of flu viruses: influenza A, B, and C. The influenza A virus has been responsible for all of the flu pandemics. There are numerous subtypes of the influenza A virus, which are separated on the basis of the identity of two glycoproteins that are found on the surface of the virus. The surface of the influenza A virus contains glycoproteins that help it invade the host's cells; they are called *hemagglutinins* (H), and they protrude from the surface of the virus looking for binding partners jutting out from the cells they want to penetrate. When the hemagglutinins find their binding partners, they latch onto them and attach to the cell, allowing the virus to invade it. Occasionally they bind to a cell, but the virus can't invade it, in which case the hemagglutinins detach and go looking for other binding sites. The influenza virus is one of only a few viruses capable of doing this. Once it has successfully bound and infiltrated the cell, the coating of the virus dissolves and the viral genes insert themselves into the host's genome. The genes are read and copied so quickly that many mistakes are made, resulting in numerous mutations and rapidly evolving viruses. As a consequence, every generation of viruses is different from its predecessors, and there is great variation among the viruses, with some being stronger and some weaker. Within 10 hours of the invasion, between 100,000 and 1 million new viruses are created from the inserted viral genes. Another glycoprotein, *neuraminidase* (N), acts like a viral machete and helps the fully formed viruses break out of the host's cells. Influenza A viruses can have 1 of 16 different hemagglutinins and 9 different neuraminidase glycoproteins. Most varieties are found only in birds, but H1, H2, H3, N1, and N2 are common in humans. The influenza A viral subtypes are identified by the species of origin, strain number, place of origin, and identity of the two glycoproteins. For example, the

2009 pandemic can be described as the 2009 H1N1 Mexican swine flu, which is quite a mouthful. Unsurprisingly, it is commonly called the swine flu or H1N1.

Although the influenza A and B viruses cause very similar symptoms, the influenza B virus can only infect humans and seals. Perhaps that is why it is responsible only for local outbreaks of the flu and there has never been an influenza B pandemic. Despite its limited transmissibility, it is estimated that influenza B is nevertheless responsible for a third of the influenza deaths in the United States.

Influenza C virus doesn't get much attention because it causes only mild upper respiratory tract illnesses and is not particularly lethal.

Because of their ability to cause pandemics, the influenza A viruses are of the most interest and concern to scientists working in the field. The H1N1 virus that was responsible for the 2009 swine flu pandemic was very efficient at moving from person to person. On March 28, 2009, Mexican health authorities reported the first indications of a new flu epidemic, and 5 months later, close to 50 million people around the world were infected, leading the World Health Organization to declare a pandemic. Research groups and health authorities all over the world sprang into action to examine the particular strain of the influenza A virus responsible for the pandemic and to create and distribute a vaccine. Fortunately, it was not a very deadly pandemic, with less than one death for every thousand individuals who were infected.

Richard Boucher and colleagues from the Cystic Fibrosis/Pulmonary Research and Treatment Center at the University of North Carolina and the Pasteur Institute in Paris are interested in the role of mucus in flu infections. The main questions they wanted to answer were whether the mucus we produce plays a protective role against inhaled pathogens, and whether it can cause airway obstruction. They created mouse models that expressed 20 times the normal amount of MUC5AC, a gel-forming mucin that is a major component of mucus, and labeled the mucus-forming protein with GFP. The mice provide the desired answers: the overproduced green fluorescent mucus was efficiently cleared from the lungs and caused no obstruction of the airways, and mucus-laden mice showed significantly lower H1N1 viral infection rates than their wild-type brethren. (4)

When the human body is infected, specialized white blood cells recognize the foreign molecules invading its cells and activate the immune system to mount a counterattack. If the infection is defeated, the immune system is primed to recognize the foreign molecules on the outside of the virus much faster and vanquish the virus if it returns, often before the symptoms of the infection are noticeable. The white blood cells don't recognize all foreign molecules; they are capable of recognizing only certain molecules, called *antigens*. The main influenza virus antigens are the neuraminidase and hemagglutinin glycoproteins, which stud the surface of the virus. In the influenza virus, the genes for these two molecules mutate faster than any other genes. This means that the neuraminidase and hemagglutinin are constantly changing to new forms that our immune systems do not recognize. To make matters worse, influenza's genetic material can be segmented, unlike most viruses where the genes are all linked in one contiguous strand of RNA or DNA. When two different influenza viruses inhabit the same cell, they can exchange segments, and completely new hybrid viruses can be formed. This increases the chance of a virus evolving in leaps and bounds. For example, both human and avian viruses can infect pigs, and if they both infect the same cell in the pig, they can swap gene fragments, thereby occasionally creating an avian flu virus that is capable of infecting humans. Extending the recipe analogy from chapter 1, in most cookbooks, the recipes are printed without interruption, and the only empty pages are found between recipes, whereas in influenza A viruses, empty pages can be inserted between segments of individual "recipes." When both a human virus and an avian virus invade a pig cell, both viruses are trying to insert their recipes into the pig's "cookbook." Occasionally the bird flu virus will insert the first segment of its instructions into the pig recipe book, and the human virus will insert the last part of its recipe. The pig cookery book now has a new recipe that is part bird and part human. Most of the time the mixed recipes are nonsense, but there are so many viruses in an infected pig, and they multiply so quickly, that even if one in a million mixed recipes resulted in an improved virus, it would be a big problem. Now that we have fluorescently tagged influenza viruses, it won't be long before scientists will be infecting individual pig cells with bird flu viruses tagged with one fluorescent color and human flu viruses tagged with fluorescent proteins of another color. This will allow them to see the interaction of the bird flu and human viruses in the pig cell mixing bowl.

Generally, as with the 2009 swine flu, the H1N1 swine virus is incredibly efficient at moving from person to person, but it's fatality rate is low. In contrast, the H5N1 avian virus is lethal when it infects humans, but fortunately it does not spread very well. Virologists and epidemiologists fear hybrid viruses, such as an infective swine flu virus with increased mortality or a deadly avian virus with improved transmissibility.

Kazuyoshi Ikuta and his colleagues at the Research Center for Infectious Diseases at Osaka University in Japan have published one of the craziest yet potentially useful applications of GFP I have ever seen. They have found a region of hemagglutinin that rarely changes in all influenza A viruses and discovered that this conserved region is recognized by the immune system. This fragment of hemagglutinin is therefore a good candidate to induce broadly reactive immunity against the influenza A virus. Because the fragment has a great deal of structural similarity with GFP, the researchers from Osaka used site-directed mutagenesis to modify GFP and a red fluorescent protein named mCherry so that part of the fluorescent protein mimicked the conserved hemagglutinin fragment while maintaining their ability to fluoresce. They found that "three of five mice immunized with each of these GFP variants followed by a booster with equivalent mCherry variants acquired anti-viral immunity against challenge with H3N2 virus at a lethal dosage." (5, p. 4981) The researchers used fluorescent proteins because they have a similar shape to the hemagglutinin fragment, and because their fluorescence is dependent on their shape, which allows the scientists to detect misfolding in their modified proteins. It is unlikely that there will ever be a vaccine consisting of a rainbow cocktail of fluorescent proteins, but the idea of using an existing protein as a scaffold has potential. Current vaccines take a long time to produce (6 months for an egg-grown virus and 2 months for a cell culture–grown virus). That is a long time when an influenza pandemic is making its way around the world. A fluorescent protein–like vaccine would be quick to produce and easy to modify against new strains. (5)

In 1918, a combination of the coldest winter recorded in the midwestern United States and the mass movement of people brought about by World War I resulted in the world's worst influenza pandemic. All over America, millions of conscripts were housed in barracks originally designed to hold just a fraction of the actual number of soldiers. Every day new soldiers joined them, and others were sent to new stations throughout the world. At the same time, workers moved to new factories built to supply the war machine. Accommodations for workers were insufficient, with space so tight that workers from one shift often shared beds with workers from other shifts. These were perfect conditions for an influenza outbreak, and it didn't take long before one came along.

It is commonly believed that the 1918 pandemic started in Camp Fulston, Kansas, where, as in most army camps at the time, the "barracks and tents were overcrowded and inadequately heated, and it was impossible to supply the men with sufficient warm clothing." (6, p. 96) The camp hospital received its first influenza victim on March 4, and within 3 weeks, 1,100 soldiers required hospitalization. Soon 10 percent of the troops in camps in Georgia had been struck, and by April, 30 of the 50 largest cities in the United States, most in close proximity to military bases, reported increased deaths due to influenza. The flu soon spread to England, France, Germany, and Spain. Spain was the only country hit by the virus that was not involved in the First World War; they were in a better position to keep accurate records and had no reason to underreport deaths therefore, it was the only country to report the true extent of the pandemic. This resulted in the mistaken belief that the 1918 flu originated in Spain and to the misnomer the Spanish flu.

The 1918 influenza pandemic struck in two waves. As mentioned earlier, the first started in the spring of 1918 and the second in the fall of that year. The second wave was unlike any other flu outbreak ever seen. The coughing and sneezing were so violent that autopsies of victims often revealed torn lungs and burst eardrums. Many patients bled from their eyes, noses, and every other orifice. The symptoms were so severe and so varied that the 1918 influenza was often misdiagnosed. Katherine Anne Porter wrote an autobiographical short novel entitled *Pale Horse, Pale Rider* about her experience with the flu, which she barely survived. She wrote, "Pain returned, a terrible compelling pain running through her veins like heavy fire, the stench of corruption filled her nostrils, the sweetish sickening smell of rotting flesh and pus; she opened her eyes and saw pale light through a coarse white cloth over her face, knew that the smell of death was in her own body, and struggled to lift her hand." (7, p. 304) The chief diagnostician of the New York City Health Department lamented, "Cases with intense pain look and act like cases of dengue. . . hemorrhage from nose. . . paralysis of either cerebral or spinal origin. . . impairment of motion may be severe or mild, permanent or temporary. . . physical and mental depression. Intense and protracted prostration led to hysteria, melancholia, and insanity with suicidal intent." (6, p. 238)

Not only were the symptoms of this strain of flu particularly severe, but they also targeted a completely different demographic than had previous outbreaks of the flu. Normally children, the elderly, and pregnant women are vulnerable to the flu, but in 1918 it was healthy adults who were most susceptible to complications and death from the flu.

More people died in a single year of the 1918 flu than were killed by a century of the Black Death. About a third of the world's population was infected and suffered from flu-like symptoms. It is estimated that between 50 and 100 million people died from the flu, which was more than 3 percent of the world's population at the time. Most of the victims were young, healthy adults. In some places the influenza sickened so many people that there was little to no medical care, and there were not enough gravediggers to keep up with the burials. Some of the Pacific Islands were particularly hard hit; for example, about 90 percent of the residents of Western

Samoa were infected, and more than 20 percent of the population died. In the 24 worst weeks of the 1918 influenza, it killed more people than AIDS has killed in 24 years. (6, p. 5)

The 1918 flu struck hard and fast. According to Major George Soper, writing in the November 8, 1918, issue of *Science*, "Like all great outbreaks of this most infectious of communicable diseases, the epidemics now occurring appear with electric suddenness, and acting like powerful, uncontrolled currents, produce violent and eccentric effects. The disease never spreads slowly and insidiously. Wherever it occurs, its presence is startling. The consternation and alarm, which it produces frequently lead to irrational and futile measures to check it." (8, p.454)

At the time of the pandemic, medical researchers did not know where this virulent and deadly strain of the flu originated and what made it different from previous seasonal flu epidemics. However, it did not escape their attention that a similar disease had been infecting pigs all over the world. It took more than 10 years after the flu had run its course until virologists isolated two closely related influenza viruses (now known as H1N1 viruses) from humans and pigs and showed that they were responsible for the 1918 pandemic. Descendants of these viruses are still found in humans and in pigs today, where they circulate in less deadly forms.

Seventy-two of the 80 residents of the Alaskan town of Brevig Mission died from the 1918 flu outbreak and were buried in the northern permafrost. One of the victims was an obese woman who was about 30 years old. The fat around her lungs prevented them from decaying, and in 2005, tissues from her frozen lungs were used to determine the complete genome of one of the viruses responsible for the 1918 pandemic. (9) This allowed researchers to suggest that the 1918 flu virus originated in birds and that sometime between 1880 and 1912 it crossed over from birds into mammals. There is some controversy about whether it first found a host in the pig population and then jumped over to humans, or if instead it made the switch to humans independently. (10) Either way, it originated from a bird flu virus. Unfortunately, none of the genes sequenced could explain the severity of the 1918 outbreak, or why this particular H1N1 virus struck young adults with such vigor. By reconstructing the 1918 virus based on its genetic sequence and then studying GFP-expressing versions of the remakes in animal models, researchers obtained partial answers to these questions. According to Darwyn Kobasa of the Canadian Public Health Agency, "One of the reasons the 1918 influenza was so severe was that infection is causing an over-stimulated immune response that actually contributes to the damage observed in the lungs of experimental animals rather than protecting the animal and assisting in clearance of the virus.

The same effect may have also played a role in the very similar extensive lung damage frequently seen in people that died during the pandemic. A similar over-stimulation of the immune response, referred to as hypercytokinemia, has also been observed in people that have died from severe disease caused by the highly pathogenic avian influenza (HPAI) H5N1 virus (or bird flu)." (11)

The bird flu virus acts differently in humans than in its native host. In birds it infects the gastrointestinal tract and not the lungs. As a consequence, bird droppings can harbor significant amounts of the avian virus and can contaminate lakes and ponds. In 1996, a new H5N1 virus was found in domestic geese in China's Guangdong province. Since then, the same avian virus has become endemic to bird populations of 6 countries and has been found in 57 other countries. Very occasionally some of them have made the transition from birds to individuals working on chicken farms. The cross-species jump is rare, and the only victims have been people with extensive contact with infected poultry. The virus has also infected other mammals such as tigers and civets. Patients infected with the H5N1 avian virus do not infect other people. Since 2003, there have been only about 600 confirmed cases of H5N1 influenza, and nearly 60 percent of the victims died from their infections. It is a deadly virus that strikes certain families harder than others, as most of the victims have been related. According to Peter Palese, a virologist at Mount Sinai School of Medicine in New York, "Infection and disease are not directly proportional to exposure. Ninety-nine point nine percent of the people who are massively exposed don't have [the] disease, and don't have antibodies in their blood. But in people who get sick, the virus replicates like crazy." (12, p. 457)

It has been impossible to eradicate the flu virus from wild birds and poultry farms or to prevent its transmission to other mammals. There is a constant cross-species transfer, and in 2005 a strain of the bird flu virus that had been circulating and mutating in mammals returned to its avian roots. It was more deadly to birds than the normal avian flu and killed more than 6,000 geese and ducks in China's largest lake before spreading to Asia, Europe, and Africa. (12)

Among virologists and epidemiologists, there is a significant fear that the H5N1 virus will mutate to a new form that is more infectious to humans. (13) Ron Fouchier, a virologist at the Erasmus Medical Center in Rotterdam, Holland, and Yoshihiro Kawaoka of the School of Veterinary Medicine at the University of Wisconsin and their colleagues have shown that it takes only five mutations to transform the H5N1 virus into a form that is transmissible from mammal to mammal by coughs and sneezes. (14, 15) Two of the five mutations have been found in wild birds, which means that they are just three mutations away from becoming transmissible between mammals. No one understands why a decade and a half of fast-paced evolution has failed to generate an H5N1 virus that can spread from person to person, but everyone is grateful for it.

The two papers describing the mutations required to make the bird flu transmissible between humans have sparked the biggest debate in science since the reports of cold fusion in 1989. Not only do they suggest that we are three mutations away from a potential H5N1 pandemic, but they have also taken us a step closer to the accidental or even intentional release of a new deadly pathogen. Both papers were accepted for publication, one in *Nature* and the other in *Science*. After 5 weeks of deliberations and more than 24 hours of conference calls, the US National Science Advisory Board for Biosecurity (NSABB) requested that the manuscripts be rewritten so that the procedure to create H5N1 viruses that are capable of spreading from person to person was not included in the papers.

Many scientists were upset that the advisory board felt the need to intervene and Fouchier himself was not convinced that the papers needed to be redacted. "For over a century, the infectious-disease community has published work on dangerous pathogens while relying on national governments, institutional biosafety offices and the responsibility of scientists to ensure that the work is done under appropriate conditions—very little has gone wrong so far, so why would that be different now?" (*16*, p.450) Many others agreed and felt that understanding host adaptation and transmissibility would lead to better detection and control of naturally occurring avian flu outbreaks. To calm the controversy and give regulators and funders an opportunity to formulate a series of best practices, 39 flu research groups from around the world agreed to stop working with mutant strains for 60 days. During the moratorium, the World Health Organization convened a panel of experts who concluded that the research should be published in full. Four months after calling for the redaction of the two papers, the NSABB reversed its decision, and the papers were published. This was not a unanimous decision, however; only 12 of the 18 NSABB members who voted were in favor of publishing the full Fouchier paper. Paul Keim, a microbiologist at Northern Arizona University in Flagstaff and acting chair of the NSABB, says, "Even the 12 who voted in favor of publication were uneasy about this uncertainty in the virus." (*17*, p. 434)

The controversy was not limited to the scientific community. It was also reported by the *New Yorker* and by *Time* magazine, which named Ron Fouchier one of its most influential 100 people of 2012.

Controlling avian flu in wild birds is an impossible task; however, limiting its spread among domesticated ducks and chickens, which are the most common bridges to human infections, should be possible. Some vaccines are available for chicken, but because there is such a wide variety of bird flu viruses and because they are constantly mutating, it is very difficult to obtain an effective one.

Genetically modified fowlpox viruses are widely used to vaccinate chickens against the bird flu, with 2 billion chickens being vaccinated with fowlpox in Mexico alone. To examine how the fowlpox virus invades a chicken cell, Michael Skinner and his colleagues, from Oxford Brookes University in the United Kingdom, have labeled a protein on the surface of the virus with green fluorescent protein and one in the chicken cell with red fluorescent protein.

"The current generation of fowlpox-based vaccines are really good at stopping chickens from getting ill when they come into contact with flu, but they don't always stop the birds from getting infected," Skinner said. "Our hope is that we can engineer a new vaccine that makes this [infection] much less likely." (18) To do this, Skinner and his collaborators developed a new technique that allowed them to see farther into cells and tissues than was possible before and to observe samples over an extended period without causing any damage. In their experiments the GFP was excited with a special infrared laser, and the green and red fluorescence was monitored as shown in figure 7.1. When the labeled viral and chicken proteins were in close contact, the green fluorescence in the virus excited the red fluorescent protein on the outside of the chicken cell. By recording

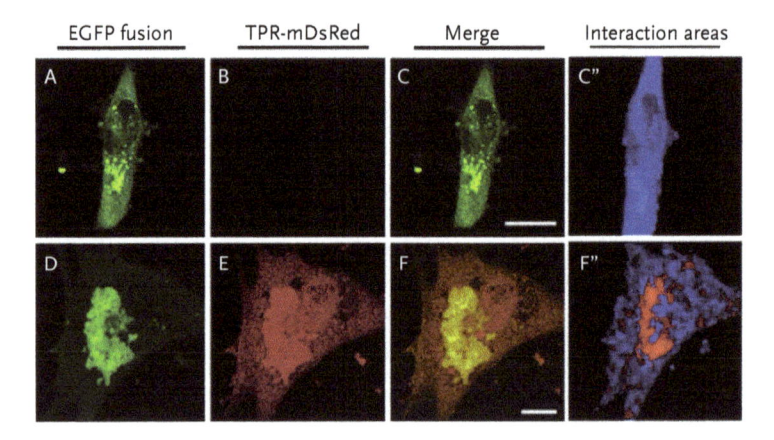

Figure 7.1 Laser scanning microscopy of chicken cells 24 hours after they were infected by fowlpox that were expressing green fluorescent protein. The top panel shows control images, in which the chicken cells were not labeled with the red fluorescent proteins. In the bottom panel, the chicken cells have a protein tagged with the red fluorescent protein DsRed. By examining the transfer of energy from the green to the red fluorescent protein (far right images), one can determine where the virus and chicken proteins are undergoing strong interactions (red) and no interactions (blue). (19)

the transfer of energy from the green to the red fluorescent protein, the researchers were able to observe how the two proteins interact. Now that they have a new technique that shows them how fowlpox and chicken proteins interact inside an inoculated chicken, they are ready to design new vaccines. (*18, 19*)

The potential for an avian flu pandemic or even just a localized outbreak has many government officials and researchers worried. Laurence Tiley, a molecular biologist at the University of Cambridge, is one of them. He has decided to take on the H5N1 virus by using genetic engineering. Together with Helen Sang, an expert in genetically modifying chickens at the Roslin Institute in Scotland, he set out to create chickens that were immune to the H5N1 virus. Although the researchers were not successful in this quest, they were able to create transgenic chickens that prevent the transmission of the bird flu virus. Their chickens can get the avian flu, but they don't pass it on to other chickens or birds.

The H5N1 virus consists of nothing but RNA and proteins. To reproduce, it must hijack a bird cell, insert its genetic information into the cell's DNA, and get the cell to make all the proteins it requires to form new viruses. One of the proteins in the virus, called *polymerase*, has the job of copying the viral RNA so that it can be converted into DNA and then be translated into proteins by the avian cellular machinery.

Polymerase recognizes the start of the viral genes it needs to copy by a short starting sequence at the beginning of each gene. Using Tiley's ideas, Sang and her students have genetically engineered chickens so that they make a short stretch of RNA that contains the starting sequences for eight genes found in the bird flu virus. In transgenic chickens infected by the bird flu virus, the viral polymerase is attracted by the transgenic RNA decoy sequences and ends up being bound before it can copy the viral genes, thereby inhibiting viral reproduction and packaging.

Tiley and Sang's chickens may prove to be a fantastic breakthrough in the fight against all viral infections. Viruses will not be able to build up a resistance against the RNA decoys because this technique is mutation proof. It interferes with the viral replication by making genetic changes that affect many of its genes, eight in this case. To overcome this, the virus would have to correctly mutate all eight genes at once and change the polymerase so that it recognizes the eight new genes at the same time.

To test whether Tiley's RNA decoy worked, Sang and her coworkers crossed some transgenic roosters having the polymerase bait with standard hens. Half of the progeny were transgenic, and half were not genetically modified. To keep track of which chickens were transgenic, the molecular biologists inserted a GFP gene into the cockerel's DNA.

All offspring with the RNA decoy binding sites also contained GFP and were fluorescent, making them easy to separate from the normal chickens (figure 7.2).

It is not surprising that Sang's group used GFP to keep track of the genetically modified chickens because her lab is also very interested in cell development. For her research in this field, she has created chickens that express GFP in every cell of their bodies. They are excellent models for what happens in human development and are much easier to work with than mice embryos. Sang says, "What we do here, and what a lot of other groups do, is take our green eggs and transplant a small number of cells from a green embryo into a normal embryo and then we can follow what happens to those cells as the embryo develops." (20) Her eggs are very popular in cell development circles, and she sends them to groups throughout the world. Dr. Seuss would be pleased to know that Sang has a colleague at the Roslin Institute who has green fluorescent pigs.

In Sang's lab ten 3-week-old transgenic chickens were infected with a high dose of the H5N1 virus. A day after they were infected, they were placed in a cage with 10 uninfected transgenic chickens. The health of all the chickens was monitored for the next 11 days. A parallel experiment was set up in which 10 nontransgenic chickens were infected and then placed in a cage with 10 uninfected chickens. The researchers found that all the directly infected birds, transgenic or not, died within 2 to 4 days after infection. This was very disappointing for the researchers, who were

Figure 7.2 The chicken on the left is normal. The one on the right has been genetically modified with GFP to differentiate it from normal chickens and with a RNA decoy sequence that prevents it from transmitting the avian influenza A virus. (Courtesy of Norrie Russell.)

hoping that their decoy RNA would prevent the virus from replicating within their transgenic chickens, thereby protecting them from the bird flu. But all was not lost because there was a clear and significant difference in the health and mortality of the initially uninfected cagemates. Seven of the chickens that had the misfortune to be placed with the infected nontransgenic chickens died within 5 days, while only two of the chickens living in the transgenic cage died during the entire 11 days. It seemed as if the transgenic chickens were not as infectious as the nontransgenic chickens.

To test this hypothesis, Tiley and his veterinary colleagues performed an experiment similar to the one described earlier, but this time they infected the chickens with one-tenth as many H5N1 viruses; placed both transgenic and nontransgenic chickens in the cage with the infected birds; and, to monitor the number of viruses being released, took daily oral and anal swabs of all the chickens. Once again the chickens caged with the infected transgenic birds had a much better survival rate even though they were not transgenic chickens themselves. The viral releases observed in the nontransgenic chickens were typical of those observed in viral infections of chickens, with oral releases preceding anal ones, whereas the infected transgenic chickens were shedding lower numbers of viruses orally and much, much lower numbers anally.

The number of viruses found in the transgenic chickens was the same as in normal chickens. Thus, the decoy didn't protect the chickens from infection, but the number of viruses released from the infected transgenic chickens was much lower than that observed in normal chickens. (*21*)

Currently, this is just a proof of concept experiment. Tiley has no intentions of selling his transgenic chickens to poultry farms in the next few years. He says, "We have more ambitious objectives in terms of getting full flu resistance before we would propose to put these chickens into true production." (*22*) There is indeed significant room for improvement, as the RNA decoy was expressed only at very low levels in these experiments, and higher levels should be attainable. Also, the RNA decoy can be supplemented with the addition of other antiviral genes. (*23*) If he can improve his technique so that the transgenic chickens cannot transmit a single virus or be immune to infection, then Tiley thinks his transgenic poultry will be ready for the market. He hopes that his RNA decoy idea may also be universally useful against infection by influenza A. Tiley explains, "Our approach is technically applicable to other domestic species that are hosts of influenza A, such as pigs, ducks, quail, and turkeys. Further development of transgenic disease resistance in poultry and other farm animals will undoubtedly stimulate debate about the application of this technology in food production." (*21*, p. 226)

For Tiley's plans to work, transgenic chickens would have to replace existing breeds where bird flu is endemic. This is theoretically possible because the global poultry industry is dominated by a handful of mega-companies. However, before the widespread adoption of transgenic chickens, the companies would have to be convinced that they could overcome consumer resistance to eating genetically modified chickens and their eggs. This will probably not occur until we have been hit by a more virulent avian flu or a more lethal swine flu pandemic.

According to Sang, the researchers have just received a new grant to continue their work with the transgenic chickens. She says they have two aims: "First, we will repeat what we have done already, and work out exactly what is happening to the flu virus in those infected transgenic chickens. And then Laurence (Tiley) has some novel ideas of additional genetic modifications to do that might confer complete resistance to the chickens." (20)

The transgenic chickens created in this research may have the best chance of any genetically modified organisms at being used to combat a disease. That is because chickens bred in the poultry industry are isolated from wild chickens and birds, and there is little chance that the genetic modifications will enter free-range chickens. Transgenic chickens are much more isolated than genetically modified plants, whose pollen blows freely in the wind, and transgenic male mosquitoes that are released so that they can fly around and find an unmodified female mates.

A question on the minds of many is whether health authorities throughout the world are prepared for a flu pandemic. In 1918, the US Army asked George Soper to investigate the flu pandemic. He found that the complete isolation of flu patients was the only way to control the outbreak. He explained, "The disease is carried from place to place by persons, not things or by the general atmosphere, as was once supposed. Its rapidity of spread is due to its great infectivity, short period of incubation, missed cases and absence of timely precautionary measures. It would appear that an epidemic does not easily start, but once the flame is well kindled a conflagration occurs which cannot be stopped. The epidemics stop themselves. This they do either by the exhaustion of the susceptible material, by a reduction in the virulence of the causative agent, or both." (8, p. 455) Soper's statement is still relevant in today's world.

What would the 1918 H1N1 virus do today? Are we ready for an avian flu virus that has undergone the few mutations required to make it more transmissible? That is a very difficult question to answer. On the one hand, the world is certainly more prepared than it was in 1918, as scientists know more about the virus, new medications are available, and some excellent surveillance systems are in place to warn us if a new flu pandemic is about to strike. The World Health Organization is in charge of one of the largest influenza surveillance networks. It actively collects and analyzes epidemiological and virological data from countries and territories around the world, and each week it publishes influenza updates online. (24) On the other hand, there are a number factors that favor the spread of the H1N1 virus through the modern world. Primary among these are increased world travel and population density, as well as the existence of intensive poultry and swine farms that can act as viral mixing bowls.

In 1980, the average person in the United States ate 45.8 pounds of chicken; by 2004, that number had nearly doubled to 84.2 pounds per person. (25) This increase in meat consumption and an ever-growing population have led to more chickens being farmed in mega-factories that can house up to 15 million birds in one facility. Large-scale pig farming has also grown dramatically, with animals housed in cramped conditions where the influenza virus can spread rapidly. The 1998 swine flu originated from one of these industrial pig farms in North Carolina. This was not surprising, says Michael Greger, director of public health at the US Humane Society. According to Greger, "North Carolina has the densest pig population in North America and boasts more than twice as many corporate swine mega-factories as any other state. With massive concentrations of farm animals within whom to mutate, these new swine flu viruses in North America seem to be on an evolutionary fast track, jumping and reassorting between species at an unprecedented rate." (26)

The flu has created an interesting dilemma for us. The pressure our civilization is placing on nature and the way we are changing nature are increasing the likelihood of an influenza outbreak. To reduce the probability of an influenza pandemic, we can either reduce our population and our meat consumption, or we can tinker with nature by creating genetically modified livestock. Either way if we want to understand and control future influenza outbreaks fluorescent proteins will be required to report on the viral schenanigans that make swine flu so infective and avian flu so lethal.

REFERENCES

1. Centers for Disease Control and Prevention. (2012). Seasonal influenza—Key facts about influenza and flu vaccine. http://www.cdc.gov/flu/pastseasons/index.htm.
2. Manicassamy, B., Manicassamy, S., Belicha-Villanueva, A., Pisanelli, G., Pulendran, B., and Garcia-Sastre, A. (2010). Analysis of in vivo dynamics of

influenza virus infection in mice using a GFP reporter virus. *Proceedings of the National Academy of Sciences of the United States of America 107*, 11531–11536.
3. Molinari, N. A. M., Ortega-Sanchez, I. R., Messonnier, M. L., Thompson, W. W., Wortley, P. M., Weintraub, E., and Bridges, C. B. (2007). The annual impact of seasonal influenza in the US: Measuring disease burden and costs. *Vaccine 25*, 5086–5096.
4. Ehre, C., Worthington, E. N., Liesman, R. M., Grubb, B. R., Barbier, D., O'Neal, W. K., Sallenave, J. M., Pickles, R. J., and Boucher, R. C. (2012). Overexpressing mouse model demonstrates the protective role of Muc5ac in the lungs. *Proceedings of the National Academy of Sciences of the United States of America 109*, 16528–16533.
5. Inoue, Y., Kubota-Koketsu, R., Yamashita, A., Nishimura, M., Ideno, S., Ono, K., Okuno, Y., and Ikuta, K. (2013). Induction of anti-influenza immunity by modified green fluorescent protein (GFP) carrying hemagglutinin-derived epitope structure. *Journal of Biological Chemistry 288*, 4981–4990.
6. Barry, J. M. (2005). *The great influenza: The epic story of the deadliest plague in history*. New York: Penguin Books.
7. Porter, K. A. (1990). *Pale horse, pale rider: Three short novels*. San Diego, CA: Harcourt Brace Jovanovich.
8. Soper, G., A. (1918). The influenza-pneumonia pandemic in the American army camps during September and October, 1918. *Science 48*, 451–456.
9. Taubenberger, J. K., and Kash, J. C. (2011). Insights on influenza pathogenesis from the grave. *Virus Research 162*, 2–7.
10. Taubenberger, J. K., and Morens, D. M. (2006). 1918 Influenza: The mother of all pandemics. *Emerging Infectious Diseases 12*, 15–22.
11. Houtekamer, C. (2009). NRC-interview met de Canadese onderzoeker Darwyn Kobasa (Oktober 2007) over het namaken van het Spaanse griepvirus. http://www.nrcnext.nl/bibliotheek/2009/08/19/
12. Yong, E. (2012). Infuenza: 5 Questions on H5N1. *Nature 486*, 456–458.
13. Fauci, A. S., and Collins, F. S. (2012). Benefits and risks of influenza research: Lessons learned. *Science 336*, 1522–1523.
14. Herfst, S., Schrauwen, E. J. A., Linster, M., Chutinimitkul, S., de Wit, E., Munster, V. J., Sorrell, E. M., Bestebroer, T. M., Burke, D. F., Smith, D. J., Rimmelzwaan, G. F., Osterhaus, A. D. M. E., and Fouchier, R. A. M. (2012). Airborne transmission of influenza A/H5N1 virus between ferrets. *Science 336*, 1534–1541.
15. Imai, M., Watanabe, T., Hatta, M., Das, S. C., Ozawa, M., Shinya, K., Zhong, G. X., Hanson, A., Katsura, H., Watanabe, S., Li, C. J., Kawakami, E., Yamada, S., Kiso, M., Suzuki, Y., Maher, E. A., Neumann, G., and Kawaoka, Y. (2012). Experimental adaptation of an influenza H5 HA confers respiratory droplet transmission to a reassortant H5 HA/H1N1 virus in ferrets. *Nature 486*, 420–428.
16. Butler, D. (2012). Freeze on mutant-flu research set to thaw. *Nature 486*, 449–450.
17. Maher, B. (2012). Bird-flu research: The biosecurity oversight. *Nature 485*, 431–434.
18. *Optics News*. (2011). Multiphoton microscopy aids bird flu fight. http://optics.org/news/2/1/14.
19. Jeshtadi, A., Burgos, P., Stubbs, C. D., Parker, A. W., King, L. A., Skinner, M. A., and Botchway, S. W. (2010). Interaction of poxvirus intracellular mature virion proteins with the TPR domain of kinesin light chain in live infected cells revealed

by two-photon-induced fluorescence resonance energy transfer fluorescence lifetime imaging microscopy. *Journal of Virology 84*, 12886–12894.
20. Sang, H. (2012). Personal communication.
21. Lyall, J., Irvine, R. M., Sherman, A., McKinley, T. J., Nunez, A., Purdie, A., Outtrim, L., Brown, I. H., Rolleston-Smith, G., Sang, H., and Tiley, L. (2011). Suppression of avian influenza transmission in genetically modified chickens. *Science 331*, 223–226.
22. Hughes, V. (2011). Transgenic chickens curb bird flu transmission. *Nature News*, http://www.nature.com/news/2011/110113/full/news.2011.16.html.
23. Timmer, J. (2011). Transgenic chickens glow green, block spread of bird flu. *Ars technica*. http://arstechnica.com/science/2011/01/transgenic-chickens-glow-green-block-spread-of-bird-flu/.
24. World Health Organization. (2012). Influenza updates. www.who.int/influenza/surveillance_monitoring/updates/en/index.html.
25. Ward, C. E. (2006). Twenty-five year meat consumption and price trends. Oklahoma Cooperative Extension Service, Oklahoma State University, Fact sheet AGEC-603.
26. Greger, M. D. (2009). CDC confirms ties to virus first Discovered in U.S. pig factories. www.humanesociety.org/news/news/2009/04/swine_flu_virus_origin_1998_042909.html.

CHAPTER 8
HIV/AIDS

	Worldwide	Sub-Saharan Africa
People living with HIV/AIDS	34 million	23.5 million (69%)
Children under 15 living with HIV/AIDS	3.3 million	3.0 million (91%)
People contracting HIV every hour	300	

In its demographics HIV has altered from an epidemic whose primary toll seemed to be within the gay communities of North America and Western Europe, to one that, overwhelmingly, burdens the heterosexual populations of Africa and the developing world. The data are so dismaying that reciting the statistics of HIV prevalence and of AIDS morbidity and mortality—the infection rates, the anticipated deaths, the numbers of orphans, the healthcare costs, the economic impact—threatens to drive off, rather than encourage, sympathetic engagement. Our imagination shrinks from the thought that these figures can represent real lives, real people, and real suffering.... I exist as a living embodiment of the iniquity of drug availability and access in Africa.... My presence here embodies the injustices of AIDS in Africa because, on a continent in which 290 million Africans survive on less than one US dollar a day, I can afford monthly medication costs of about US$400 per month. Amidst the poverty of Africa, I stand before you because I am able to purchase health and vigour. I am here because I can afford to pay for life itself.

Edwin Cameron, *justice of the Constitutional Court of South Africa*,
XIIIth International AIDS Conference, Durban, July 10, 2000

There is no other disease that has caused the same amount of global suffering and panic as has acquired immune deficiency syndrome (AIDS) in its short existence. The disease is caused by the human immunodeficiency virus (HIV), which compromises the human immune system, resulting in an increased susceptibility to infections from bacteria, fungi, viruses, and parasites.

HIV is thought to have arisen from the simian immunodeficiency virus (SIV), found in chimpanzees. SIV was most likely transmitted to humans and mutated to HIV when they hunted and ate infected chimpanzees. Hunters, whether human or animal, are very susceptible to interspecies viral infections. Catching an animal like a chimpanzee and then preparing it for consumption involves many opportunities for fluid exchange between the hunter and the victim. Beatrice Hahn and Martine Peeters, (1) whom we met in chapter 4 because their groups were responsible for the theory that *Plasmodium falciparum* came to humans from wild gorillas, have shown that chimpanzee SIV itself came from a combination of viruses that inhabited red-capped mangabey and white-nosed guenon monkeys. Chimpanzees find both these species of monkeys very tasty, and it is not difficult to imagine a chimpanzee infected with SIV from one of these species eating an SIV-infected monkey of the second species. The two viruses could meet in one of the chimpanzee's cells and exchange genetic material, making a new SIV that would ultimately cross over to humans. Genetic studies suggest a number of animal-to-human crossings, with the most common recent ancestor dating back to 1910.

In *The Viral Storm*, Nathan Wolfe describes witnessing some chimpanzees eating monkeys they had caught: it was "an instant visual example of the blurring of lines between species. The manner in which they were ingesting and spreading fresh blood and organs was creating the ideal environment for any infectious agents present in the monkeys to spread to the chimpanzees. The blood, saliva, and feces were splattering into the orifices of their bodies (eyes, noses, mouths, as well as any open sores or cuts on their bodies)—providing the perfect opportunity for direct entry of a virus into their bodies." (2, p. 43)

The first proven cases of AIDS date back to 1959, when there was a localized outbreak in Leopoldville, which was then part of the Belgian Congo and is now Kinshasa in the Democratic Republic of the Congo. (3) During the outbreak, the lymph node of a woman who died from the disease, which was unknown at the time, was embedded in wax. Forty-five years later, an evolutionary virologist, Michael Worobey, and his colleagues were able to find genetic evidence of an HIV infection and deduced the disease had left its primate hosts sometime between 1902 and 1921. (4)

It took more than 50 years from the time it left its primate hosts before HIV started causing large-scale havoc among the human population. In Zaire in the mid-1970s, AIDS appeared as a devastating disease that resulted in victims wasting away. No one was able to identify the disease, which was known locally as "Slim"; subsequently, it has been shown to be AIDS. No preventive measures were taken, and it slowly but steadily spread across the world. In June 1981, AIDS was first recognized as a new disease that was responsible for an inordinately high number of young gay men in New York and Los Angeles getting lung infections and a rare form of cancer, Kaposi's sarcoma, that was usually only seen in older patients. By the end of that year, AIDS cases were reported from the United Kingdom and among intravenous drug users. A few months later it was clear that hemophiliacs were also vulnerable to the disease, which was officially named AIDS in August 1982. The blood transfusions of hemophiliacs can expose them to the blood of several thousand donors, which led researchers to suspect that the disease is transmitted in the blood of AIDS patients.

Finding the cause of AIDS was a significant medical breakthrough, but because of the many political and social aspects of the disease, it was controversial. In the May 1983 issue of *Science*, Luc Montagnier and Françoise Barré-Sinoussi, two medical researchers at the Pasteur Institute in Paris, announced that they had isolated a new virus that might be responsible for AIDS. In the same month they presented their findings at the annual Cold Spring Harbor meeting, at the Centers for Disease Control, Long Island, and the National Institutes of Health in Bethesda, Maryland. The Pasteur Institute researchers applied for patents to use the virus as a diagnostic tool for AIDS and sent the virus, which they named lymphadenopathy-associated virus (LAV) to the Centers for Disease Control and the National Cancer Institute in the United States. Despite its importance and its publication in *Science*, the discovery was not covered by the news media and did not make waves among AIDS researchers.

Nearly a year later, in a carefully orchestrated media event, US Health and Human Services secretary Margaret Heckler announced that Robert Gallo of the National Cancer Institute had isolated the virus that causes AIDS and that there would soon be a commercially available test to detect the virus with "essentially 100 percent certainty." Heckler was correct. It didn't take long before there was an AIDS test that could detect the presence of antibodies to the AIDS virus. By the time the test was available, however, it had been established that Gallo's virus was the same virus that Montagnier had sent to the National Cancer Institute and that Gallo's work was an extension of Montagnier's research, not a new paradigm-shifting discovery. This led to the Pasteur Institute filing a lawsuit against the National Cancer Institute for a share of the royalties from its patented AIDS test. The stakes were high, and there was even a disagreement about what to name the virus that was responsible for

AIDS. According to the May 12, 1986, issue of *Time*, "The name of the virus had itself become a political football as the French insisted on LAV (lymphadenopathy-associated virus), while Gallo's group used HTLV-3 (human T cell lymphotropic virus, type 3)." (5, p. 86) In the end it was the International Committee on the Taxonomy of Viruses that resolved the issue by ruling that from June 1986 onward the virus should be called the human immunodeficiency virus (HIV). (6) The resolution came just in time for the naming of a new related virus that was isolated from patients in West Africa and was now appropriately named the HIV-2 virus.

The naming controversy was over, but the Pasteur Institute lawsuit was still pending, and the legitimacy of Gallo's claim to the discovery of HIV and the link between HIV and AIDS were still disputed. After substantial pressure from the Reagan administration, on March 31, 1987, at a ceremony attended by the president it was announced that the Pasteur Institute would drop its lawsuit and the profits from the HIV antibody test would be shared. Diplomatic pressure at the highest levels had seemingly settled the issue.

But after a 20-year hiatus, the HIV discovery row was revived when the Nobel Foundation awarded the 2008 Nobel Prize in Physiology or Medicine to Harald zur Hausen, Luc Montagnier, and Françoise Barré-Sinoussi. Half of the prize was awarded to Harald zur Hausen "for his discovery of human papilloma viruses causing cervical cancer," and the other half went jointly to Françoise Barré-Sinoussi and Luc Montagnier "for their discovery of human immunodeficiency virus." The Nobel Prize can be awarded to no more than three researchers. The foundation could easily have awarded the prize to Montagnier, Barré-Sinoussi, and Gallo, but it pointedly split the award in half so that the prize to went to zur Hausen, thereby using up the third slot. In awarding the prize, the chair of the Nobel Committee, Bertil Fredholm, stated: "I think it is really well established that the initial discovery of the virus was in the Institute Pasteur." (7) To this day, Gallo and the US government claim the Gallo lab was the first to conclusively show a link between HIV and AIDS, and that researchers in Gallo's laboratory independently found HIV; the similarity between their virus and Montagnier's virus was due to an accidental contamination of their work with Montagnier's virus. (Notably, 2008 was the same year the Nobel Prize in Chemistry was awarded to Shimomura, Chalfie, and Tsien for their discovery and development of the green fluorescent protein.)

HIV is a spherical retrovirus with a diameter that is approximately 60 times larger than a blood cell, which is large for a virus and much larger than the influenza or dengue viruses. There are typically 72 envelope proteins that stick out of the virus; at their very tips the envelope proteins have some sugar molecules called *glycoproteins*. These sugars are the key that HIV uses to get into the macrophage and CD4⁺ T cells of an infected individual. Cells require energy to function; sugars are an excellent source of this energy. The cell membranes of the macrophage and CD4⁺ T cells have receptor sites on their surface that recognize sugar molecules. When a sugar molecule comes along, the cell membrane opens up and ingests the "food." The receptor molecules are easily fooled by the sugar-coated HIV surface and open up expecting nothing more than a few sugar molecules. HIV takes advantage of this trickery by spewing its guts, the bullet-shaped HIV capsid, into the cell. The damage is done, for the HIV capsid contains all the tools the virus needs to hijack the invaded cell and convert it to a HIV viral factory. Inside the capsid are two single strands of HIV RNA, both having a complete copy of all the virus's genes and three replication enzymes that will subvert the host cell's machinery to copy the viral RNA into DNA and then produce new HIV in the host cells (figure 8.1).

One of these enzymes, reverse transcriptase, releases the RNA from the plasmid and rather sloppily copies it into a DNA strand. The mutations resulting from the errors made by the reverse transcriptase are fortuitous for the virus, because they are what make future generations of the virus different and are responsible for its drug resistance and ability to evade the body's immune system. The polymerase from the viral capsid converts the heavily mutated DNA strand into a double helix that is indistinguishable from the host cell DNA. In the nucleus of the invaded cell, the aptly named viral enzyme, integrase, sneaks the DNA with the viral message into the host's DNA. Satisfied that the damage is done, a time bomb has been planted in the nucleus, the viral tools that were smuggled inside the plasmid rest; this is the dormant stage of the HIV infection. CD4⁺ T cells are white blood cells that play a central role in cell-mediated immunity by being activated to fight infections. HIV has evolved to strike when the host defenses are at their weakest. The DNA bomb, hidden among the host's genome, explodes when the T cells are activated to combat an infection. The DNA is converted back into RNA, leaves the nucleus, and is packaged into a new capsid, and a mature virus is produced that buds out of the cell ready to replicate in the next cell. A single cell can be invaded by numerous HIV strains, producing new virons with RNA sequences from all viruses, resulting in increased genetic variability. Ten billion virons can be generated every day; this speedy reproduction leads to more mutations and even greater genetic diversity among HIV virons. The HIV replication destroys the infected cells and eventually leads to disease progression (see figure 8.1).

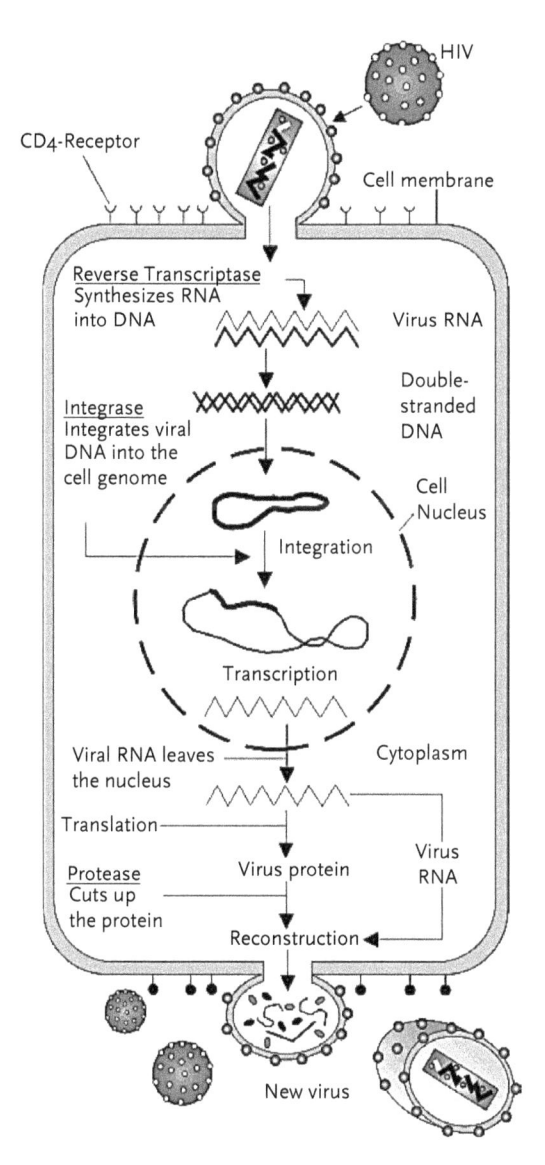

Figure 8.1 Schematic summarizing the text and illustrating how the HIV enters a CD4⁺ cell, integrates the viral DNA into the cell's genome, and finally re-emerges as a new HIV.

HIV is an exception among the retroviruses; it is the only retrovirus that destroys the cells it infects. This is incredibly important in understanding and studying AIDS because it is the wholesale slaughter of activated CD4⁺ T cells that results in the severe immunodeficiency that is characteristic of AIDS. In 2013, Gary Nabel and his coworkers at the Virology Laboratory in the National Institute for Allergy and Infectious Diseases of the National Institutes of Health in Bethesda, Maryland, used GFP to show how the virus causes cell death. Mammalian cells have evolved intricate protection mechanisms against DNA damage. Normally, when a retrovirus inserts its (viral) DNA into the host cell's genome, DNA repair enzymes are activated to clean up the mess. In activated CD4$^+$ T cells infected with HIV, integration of viral DNA also leads to expression of DNA repair enzymes; however, one of these repair enzymes has gone bad, and instead of repairing the DNA, it promotes programmed cell death. Hopefully, this finding by Nabel and colleagues will lead to the development of new drugs that inhibit the rogue DNA repair enzyme, thereby prolonging CD4$^+$ cell survival. (8)

One of the great difficulties in curing a patient with HIV is that the virus undergoes long periods of latency and only becomes active once the white blood cells are required to fight an infection. It used to be very difficult to recognize and isolate cells infected with the dormant form of HIV, but now Eric Verdin and his colleagues at the Gladstone Institute of Virology and Immunology, San Francisco, have created HIV reporters that are green (GFP) in the active form and red (mCherry) in the latent form. According to the authors, "These reporters are sufficiently sensitive and robust for high-throughput screening to identify drugs that reactivate latent HIV. These reporters can be used in primary CD4$^+$ T lymphocytes and reveal a rare population of latently infected cells responsive to physiological stimuli. . . . our HIV-1 reporters. . . open up new perspectives for studies of latent HIV infection." (9, p. 283)

HIV drugs are designed to stop the prolific HIV reproduction that occurs in the CD4$^+$ T cells. Fusion inhibitors prevent the virus from binding to the CD4$^+$ receptors, thereby barring entry of HIV into the cell. Some antiretroviral drugs prevent the viral reverse transcriptase from copying the viral RNA in the host's cell. Numerous drugs block transcription in different ways. Proteases that form the DNA helix can be inhibited, and integrase can be prevented from splicing viral DNA into the host's DNA. The combination of a variety of these antiretroviral drugs is very effective, and so-called drug cocktails are the current treatment of choice.

One of the questions that has plagued HIV researchers is how the newly released HIV cells can find new CD4$^+$ T cells to hijack so quickly. The answer has been found: When human T cells bump into each other, they form a sticky strand that connects the two cells. These strands, dubbed "membrane nanotubes" by Daniel Davis and his Imperial College colleagues who discovered them, can connect two T cells that are several cell lengths apart. By infecting a T cell with HIV containing GFP-labeled proteins, the researchers were able to show that HIV proteins travel down

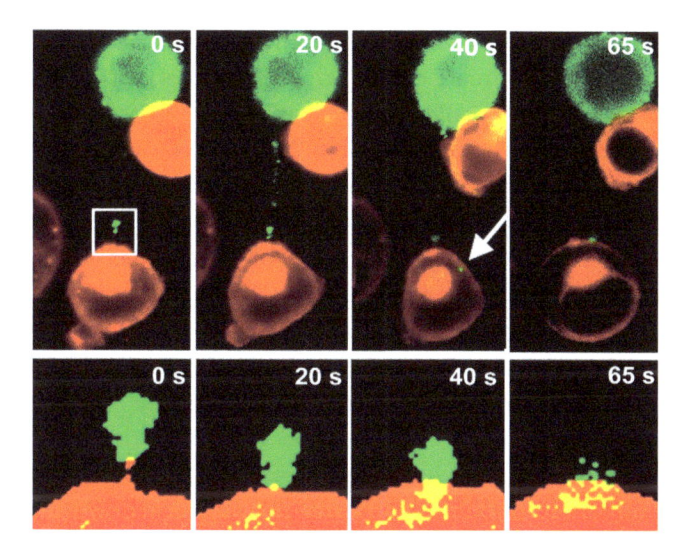

Figure 8.2 Time-lapse imaging of GFP-tagged proteins moving along a membrane nanotube connecting infected with uninfected T cells. The boxed regions are enlarged to show that the protein reaches the initially uninfected T cell. (*10*)

the nanotubes from infected to noninfected cells (figure 8.2). These nanotubes may be part of the reason HIV is so effective at spreading rapidly within host bodies and may explain why in vitro cell-to-cell HIV infection is a thousandfold more efficient than with cell-free virus. (*10*) According to the authors, "Our data show that T-cell nanotubes are novel physical connections between T cells that can have important consequences for allowing rapid spread of HIV-1. As the ability of HIV-1 to spread between cells is a major determinant of its virulence, this mechanism of HIV-1 transmission could be important to its pathogenicity and may open new avenues for drug targets." (*10*, p. 218)

The experiments shown in figure 8.2 were done in cell cultures. In 2012, Thorsten Mempel, a professor at the Center for Immunology and Inflammatory Diseases, Massachusetts General Hospital, Harvard Medical School in Boston, went the next step and used fluorescent HIV to spy on HIV 1 infected T cells in the lymph nodes of live humanized mice. The work of his laboratory showed that the viruses use $CD4^+$ T cells to travel through the body and infect other $CD4^+$ T cells. (*11*) According to Mempel, "We have found that HIV disseminates in the body of an infected individual by 'hitching a ride' on the T cells it infects. Infected T cells continue doing what they usually do, migrating within and between tissues such as lymph nodes, and in doing so they carry HIV to remote locations that free viruses could not reach as easily." (*12*)

The proliferation of HIV in infected cells destroys the cells or impairs their function. The ultimate consequence of the destruction of the cells, particularly the CD4$^+$ T cells, is the onset of AIDS. Once the number of CD4$^+$ T cells drops from approximately 1,200 cells per microliter of blood to below 200 cells, cellular immunity is lost. AIDS is defined by the presence of certain infections that are not typically found in people with healthy immune systems and a CD4$^+$ T cell count that has dropped below 200 cells per microliter. Without any treatment, an HIV infection progresses into AIDS in 9 to 10 years; after the onset of AIDS, the median survival time is only 9.2 months. Although it takes many years for the HIV infection to exhibit all the symptoms associated with AIDS, the majority of CD4$^+$ T cells are destroyed in the first few weeks of the infection. Infection can occur when bodily fluids such as blood, semen, vaginal fluid, pre-ejaculate, or breast milk that contain HIV in a free-floating form or within infected immune cells are transferred from person to person.

There is no cure or vaccine for AIDS/HIV despite the fact that since 1989 more than 50 potential vaccines have been studied using more than 30,000 healthy volunteers. Currently, the most efficient treatment available is triple antiretroviral therapy. The so-called cocktails of highly active antiretroviral agents have been used since 1996 and have changed AIDS from a fatal disease to a long-term treatable condition. More than 30 antiretroviral drugs and fixed-dose drug combinations are now available. Treatments have resulted in remarkable improvements in the quality of life for patients and a notable decrease in HIV-associated mortality. However, triple antiretroviral therapy does not cure HIV infections, and cessation of the treatment results in the re-emergence of the HIV infection. It has changed the face of AIDS in the countries that can afford to prescribe the cost of the cocktails, which is roughly $20,000 to treat one patient for a year. Thanks to large amounts of international funding, 6 million people infected with HIV are being treated with antiretrovirals in the developing world. However, the United Nations Program on HIV/AIDS shows that there are approximately 10 million more people who need the treatment. Recent studies have shown that the spread of AIDS can be controlled if antiretroviral drugs are used very early in infections, or even in uninfected people in high-risk groups. For example, a report by the US National Institutes of Health shows that when patients are treated early, a reduction of 96 percent in infection to HIV-free partners is observed.

Antiretroviral drugs have certainly improved the lives of HIV-infected patients who can afford their high cost. However, it is unlikely that they will control the spread of AIDS, and so the search for an AIDS vaccine or alternative prevention methods continues. One potential method being investigated is gene therapy. Cats are an important model system in the study of AIDS and in proof of concept studies for the development of an HIV gene therapy. Cats are studied because feline immunodefiency virus (FIV) is a very close relative to HIV, both in its genomic structure and in the way it attacks its host. In both species the virus causes AIDS by depleting the infection-combating T cells. Studies of FIV are not only important for modeling HIV but also are valuable in their own right because millions of cats contract and die of FIV/AIDS every year.

Eric Poeschla and coworkers at the Mayo Clinic College of Medicine, Rochester, Minnesota and in Yamaguchi University, Japan, have introduced two new genes into cats—a gene for a protein that is known to inhibit HIV-1 and FIV replication in rhesus monkeys, and GFP (figure 8.3). (13) The green fluorescence was used to show that the genetic modifications were successful and to quantify how much of the protein was being produced. The transgenic cats and their offspring all express GFP and the restriction factors from the rhesus monkey throughout their bodies. The work garnered a significant amount of media attention; after all, it is very difficult to resist writing about and showing photographs of green fluorescent cats. Most of the articles focused on the green fluorescence and ignored the HIV resistance results, which, although tentative, are very positive. Poeschla and his colleagues cultured white blood cells from the genetically modified cats and found FIV replication was inhibited in these cells. "At this point, we don't know whether the whole animal is protected from acquisition of the virus or development of the associated illness," said Poeschla. "Whole-body animal testing will have to be done." (14)

Figure 8.3 This kitten has been genetically modified to express a restriction factor from a rhesus monkey and GFP. The experiments were a success; FIV, the feline form of HIV, does not replicate in white blood cells obtained from the kittens. (13)

As you can probably imagine, fluorescent proteins are incredibly useful for tracking HIV virons in cell cultures and in live organisms, and not surprisingly, this is the most common application for fluorescent proteins in AIDS research. But scientists are always pushing the limits, including the diffraction limit, which prevents microscopists from recording the interaction of an human immunodeficiency virus with cellular proteins during the entry and assembly stages. Now super-resolution microscopy techniques allow for nanometer resolution of fluorescently labeled molecules in intact cells, and we can observe molecular interactions occurring at a scale smaller than is theoretically possible. (15) Fluorescent proteins are illuminating every aspect of HIV's existence, and every year we can image its behavior deeper in living organisms and with more resolution than ever before.

REFERENCES

1. Bailes, E., Gao, F., Bibollet-Ruche, F., Courgnaud, V., Peeters, M., Marx, P. A., Hahn, B. H., and Sharp, P. M. (2003). Hybrid origin of SIV in chimpanzees. *Science 300*, 1713.
2. Wolfe, N. (2011) *The viral storm: The dawn of a new pandemic age*. New York: Times Books.
3. Zhu, T., Korber, B. T., Nahmias, A. J., Hooper, E., Sharp, P. M., and Ho, D. D. (1998). An African HIV-1 sequence from 1959 and implications for the origin of the epidemic. *Nature 391*, 594–597.
4. Worobey, M., Gemmel, M., Teuwen, D. E., Haselkorn, T., Kunstman, K., Bunce, M., Muyembe, J.-J., Kabongo, J.-M. M., Kalengayi, R. M., Van Marck, E., Gilbert, M. T. P., and Wolinsky, S. M. (2008). Direct evidence of extensive diversity of HIV-1 in Kinshasa by 1960. *Nature 455*, 661–664.
5. Willis, C. (1986). A different kind of AIDS fight. *Time*, May 12, 86.
6. Coffin, J., Haase, A., Levy, J. A., Montagnier, L., Oroszlan, S., Teich, N., Temin, H., Toyoshima, K., Varmus, H., Vogt, P., and Weiss, R. (1986). New name for AIDS virus—What to call the AIDS virus. *Nature 321*, 10.
7. *The Guardian* (2008, 7 October). Nobel awards revive HIV discovery row.
8. Cooper, A., Garcia, M., Petrovas, C., Yamamoto, T., Koup, R. A., and Nabel, G. J. (2013). HIV-1 causes CD4 cell death through DNA-dependent protein kinase during viral integration. *Nature 498*, 376–379.
9. Calvanese, V., Chavez, L., Laurent, T., Ding, S., and Verdin, E. (2013). Dual-color HIV reporters trace a population of latently infected cells and enable their purification. *Virology 446*, 283–292.
10. Sowinski, S., Jolly, C., Berninghausen, O., Purbhoo, M. A., Chauveau, A., Kohler, K., Oddos, S., Eissmann, P., Brodsky, F. M., Hopkins, C., Onfelt, B., Sattentau, Q., and Davis, D. M. (2008). Membrane nanotubes physically connect T cells over long distances presenting a novel route for HIV-1 transmission. *Nature Cell Biology 10*, 211–219.

11. Murooka, T. T., Deruaz, M., Marangoni, F., Vrbanac, V. D., Seung, E., von Andrian, U. H., Tager, A. M., Luster, A. D., and Mempel, T. R. (2012). HIV-infected T cells are migratory vehicles for viral dissemination. *Nature 490*, 283–287.
12. McGreevey, S. (2012). HIV-infected T cells help transport the virus throughout the body. News release. Massachusetts General Hospital. www.massgeneral.org/about/pressrelease.aspx?id=1491.
13. Wongsrikeao, P., Saenz, D., Rinkoski, T., Otoi, T., and Poeschla, E. (2011). Antiviral restriction factor transgenesis in the domestic cat. *Nature Methods 8*, 853–859.
14. Temple, N. (2011). Glow-in-the-dark transgenic cats. *Cosmos—The Science of Everything*. www.cosmosmagazine.com/news/glowing-transgenic-cats-resist-feline-hiv/.
15. Pereira, C. F., Rossy, J., Owen, D. M., Mak, J., and Gaus, K. (2012). HIV taken by STORM: Super-resolution fluorescence microscopy of a viral infection. *Virology Journal 9*, 84–90.

CHAPTER 9

Diseases of the Brain

Brain disease	Number of people currently living with the disorder in the United States
Alzheimer's	4,000,000
Parkinson's	1,000,000
Huntington's	30,000
ALS	25,000

And men ought to know that from nothing else but thence [from the brain] come joys, delights, laughter and sports, and sorrows, griefs, despondency, and lamentations. And by this, in an especial manner, we acquire wisdom and knowledge, and see and hear, and know what are foul and what are fair, what are bad and what are good, what are sweet, and what unsavory.. . . And by the same organ we become mad and delirious, and fears and terrors assail us.. . . All these things we endure from the brain, when it is not healthy.

Hippocrates, *The Genuine Works of Hippocrates*

The man whom many consider the father of modern neuroscience laid the foundations for future experiments that would blur the line between reality and science fiction. Perhaps it is appropriate that he also dabbled in science fiction, writing under the pseudonym "Dr. Bacteria." Thanks to "Dr. Bacteria's" scientific work, mind-reading and mind control are subjects that are no longer limited to science fiction. Today they appear more often in scientific papers than in science fiction stories. Fortunately, they are not being used for nefarious purposes; instead, their development has been driven by a need to understand and cure neurodegenerative diseases.

"Dr. Bacteria," or Santiago Ramón y Cajal, as he was known to most, was born on May 1, 1852, in Petilla de Aragón, Spain. His autocratic father was a professor of applied anatomy who was adamant that his son would study medicine. However, Cajal was a rebellious and difficult youth who was more interested in drawing and pulling pranks than studying medicine. At age 11, he built his own cannon, loaded it with rocks, and shot down the neighbor's gate. As a result, Cajal was sent to jail for 3 days. Undeterred, he built a bigger cannon, which blew up in his face and nearly blinded him. His father sent him to a number of schools in the hope that he would see the light and turn to medicine, but Cajal was unconvinced. Why, he asked, "exchange the magic palette of the painter for the nasty and prosaic bag of surgical instruments! The enchanted brush, the creator of life. . . . Be given up for the cruel scalpel, which wards off death." (1, p. 102)

Despite Santiago Ramón y Cajal's passion for painting, his father prevailed, and after completing high school Santiago began studying medicine under the guidance of his father at the University of Saragossa. Upon qualifying, he joined the army as a doctor and was posted to Cuba from 1874 to 1875. While there, he contracted both malaria and tuberculosis. Despite his expertise with cannons, the army was not Cajal's thing, so he returned to Spain, settled down in academia, got married, and fathered four daughters and three sons.

The microscope was Cajal's new weapon of choice; he was going to use it to study the brain. The human brain has approximately 10 billion nerve cells with a trillion connections laid out in an indecipherable tangle. Figuring out what's what in the brain was both a daunting and a tedious task when Cajal started, especially if one considers the microscopes available to researchers at the time, which were not much better than a modern magnifying glass. This was a field that required someone with perseverance and tenacity, someone who was jailed for 3 days at the tender age of 11 and bounced right back with a bigger bang, someone who would sit in front of a microscope for hours on end without being discouraged. Cajal's choice of neurohistology as his area of specialization made good sense in many other respects as well, for it also required someone with the artistic skill to record what he was seeing under the microscope. Furthermore, after the initial purchase of a microscope, the laboratory costs were low, which was important to Cajal, who did not have much funding for his research.

During Cajal's lifetime, histology was an art that relied on staining cells so that they were distinguishable from each other and from their background. In fact, in 1902 Gustav Mann, a pioneer in histological techniques, wrote that "to be an histologist became practically synonymous with being a dyer, with this difference—the professional dyer knew what he was about, while the histologist with a few exceptions did not know, nor does he to this day." (2, p. 190) In 1873, Camillo Golgi developed the stain that would untangle some of the neural mess for Cajal. A resident in a chronic disease hospital in Pavia, Italy, Golgi combined potassium dichromate and silver nitrate. In brain cells the potassium dichromate reduced the silver nitrate to black metallic silver, staining the entire cell black, in magnificent contrast against the yellow background of the chromate-stained brain; however, the stain was so unreliable that it had already started sliding into disuse when Cajal first saw the technique about 14 years after it was developed. Despite the drawbacks, he was drawn to the Golgi method because it stained only a few of the cells, allowing them to stand out among all the unstained cells, and because cells traversing a cube of brain matter could be stained. Nerve cells appeared "colored brownish black even to their finest branchlets, standing out with unsurpassable clarity upon a transparent yellow background. All was sharp as a sketch with Chinese ink," Cajal wrote in his autobiography. (3, p. 306)

After much fiddling and tweaking, Cajal improved the stain and finally got reproducible results, especially if he avoided heavily myelinated cells. Fetal and neonatal brains of birds and small mammals, in which the myelination is less dense and sometimes completely absent, became the object of his obsession. Even with the improved Golgi stain, it was not easy to trace the path of a brain cell in a cube of brain matter. The difficulty with viewing brain slices under the microscope was that the stained brain cells were never in the plane of the slide, and so Cajal would have to view a series of brain slices and remember where the stained cell intersected all the slides before composing a mental picture of the path of the brain cell through the slides. Thanks to his fantastic memory and feel for space, Cajal was also able to generalize neural structures from the hundreds of birds and small mammals he had examined.

Today, neuroscientists still admire the accuracy of Cajal's paintings, more than a hundred years after they were drawn (figure 9.1). But it is not the accuracy of his drawings that is responsible for Cajal being named the father of neuroscience; it is the fact that his drawings proved that neurons were autonomous units, with spaces between adjacent brain cells, that has led to his fame. His work was the final nail in the coffin for those who believed that all brain cells were connected to each other and that the brain acted as a whole in producing its effects. Another impressive finding of Cajal's that was to stand the test of time was his discovery that "nervous current" has directionality—a conclusion he came to without any knowledge about ion channels and action potentials. (4)

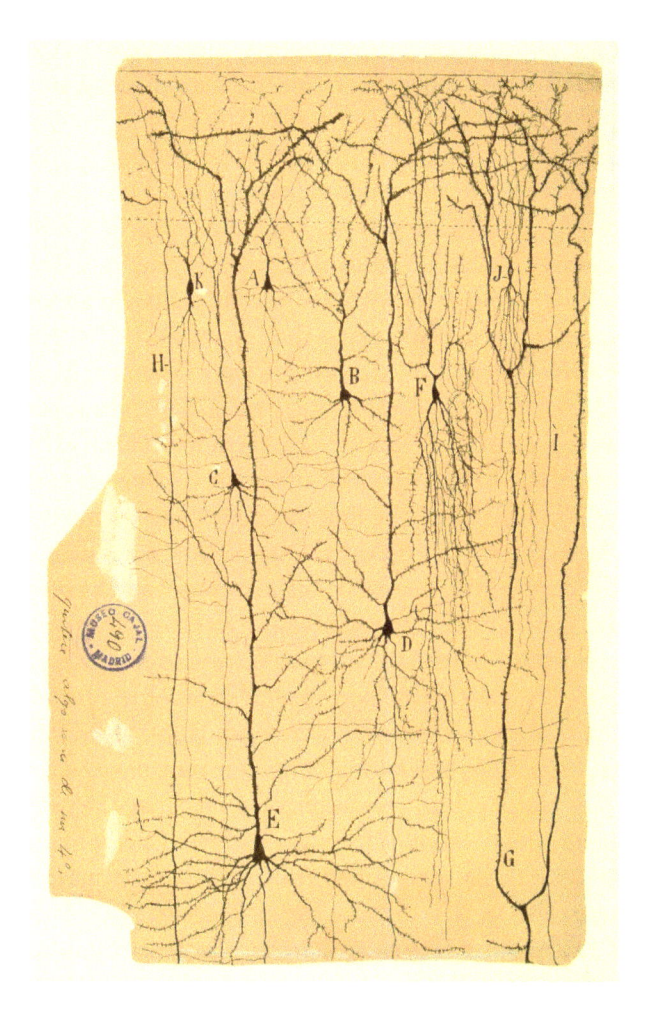

Figure 9.1 Drawing of the neocortex, the outermost layer that forms the folds of the brain, by Santiago Ramón y Cajal, 1899. (Santiago Ramón y Cajal. Cajal Legacy. Instituto Cajal, CSIC. Madrid.)

In 1906, Cajal and Golgi were jointly awarded the fifth Nobel Prize in Physiology or Medicine "in recognition of their work on the structure of the nervous system." (5) The interactions between Cajal and Golgi during the Nobel awards week would have been interesting to watch as Golgi was still an ardent believer in the theory of a single unified neural network, a belief he would hold until his death. This was the same network that he advocated in his Nobel acceptance speech and that Cajal ridiculed in his reply by saying, "True, it would be very convenient and very economical from the point of view of analytical effort if all the nerve centres were made up of a continuous intermediary network between the motor nerves and the sensitive and sensory nerves. Unfortunately, nature seems unaware of our intellectual need for convenience and unity, and very often takes delight in complication and diversity." (6, p. 240)

If only Cajal could time-travel, it would be fascinating to see what Cajal the artist and neuroscientist would think of modern imaging techniques. Never before has the brain been so beautiful, especially not since Jeff Lichtman and Joshua Sanes, researchers at the Harvard Brain Center, have created transgenic mice with fluorescent multicolored neurons. The photographs of their mouse brains that appeared in the November 1, 2007, issue of *Nature* could be displayed in the Museum of Modern Art.

The mice, created by a genetic strategy termed *Brainbow*, may have the same effect on neuroscience that Google Earth has had on cartography. It extends the work of Cajal from black and white to color, and takes it from two to three dimensions. In the Brainbow mice, the Harvard researchers have introduced genetic machinery that randomly mixes green, cyan, and yellow fluorescent proteins in individual neurons, thereby creating a palette of 90 distinctive hues. "The technique drives the cell to switch on fluorescent protein genes in neurons more or less at random," says Jean Livet, the postdoctoral researcher responsible for most of the laboratory work that resulted in the Brainbow mice. "You can think of Brainbow almost like a slot machine in its generation of random outcomes." (7)

Using a rainbow of colors, researchers can now map the neural circuits of the brain. (8) Golgi's stain selectively blackened a few neurons, allowing Cajal to distinguish the stained neurons from the mass of neurons surrounding them. In Brainbow mice the individually colored neurons help define the complex tangle of neurons that constitute the brain and nervous system (figures 9.2 and 9.3). And thanks to modern computer technology, users of Brainbow don't need to have Cajal's prodigious memory; they don't need to remember a sequence of slides to visualize the three-dimensional wanderings of a neuron. A cube of transgenically modified Brainbow cerebral matter can be sliced into hundreds of thin slivers, which are photographed and recorded so that the images of all the slices can be computationally recompiled to produce a three-dimensional representation of the cube.

In a 2011 profile of Joshua Sanes, *The Scientist* asked him how techniques like Brainbow are useful. "They're really good for getting pictures on the covers of journals and books," he joked before describing how his technique allows the average neuroscientist to do what Cajal did more than a century ago. "Seriously, they allow you to distinguish individual neurons out of a morass. Just like Cajal did with his Golgi stain. If you stain a single neuron in its entirety you can learn about its shape and its connections. If

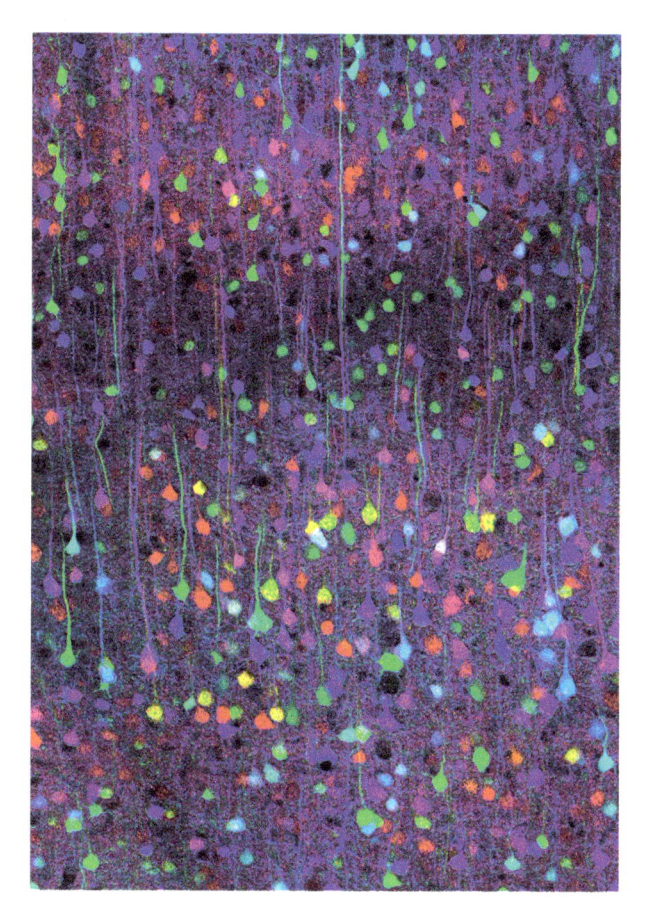

Figure 9.2 Mouse cerebral cortex imaged with Brainbow (8) (Confocal image courtesy of Tamily Weissman; mouse courtesy of Jean Liven and Ryan Draft.)

you label 100 neurons in a single mouse, you learn much more than if you study 100 mice with one neuron labeled in each. Cajal was an incredible genius in that he could look at one neuron in each of 100 mice and then go home and draw a picture that synthesized all of that information. And almost always he got that right. Anyone else would have gotten it wrong. Because you can't register all those variations against each other unless you are supernaturally brilliant like Cajal." (9, p. 55) That is why you need Brainbow or its newest version, Brainbow3.

In 2011, Atsushi Miyawaki reported a treatment that renders brain tissue transparent (figure 9.4). (10) The technique, Scale2, which takes 2 weeks and doesn't interfere with internal fluorescence, allows researchers to observe fluorescent neurons five times deeper than before. Miyawaki has yet to use Scale2 on Brainbow-modified mice, but he foresees no problems. The biggest limitation seen by Miyawaki at present is the need to

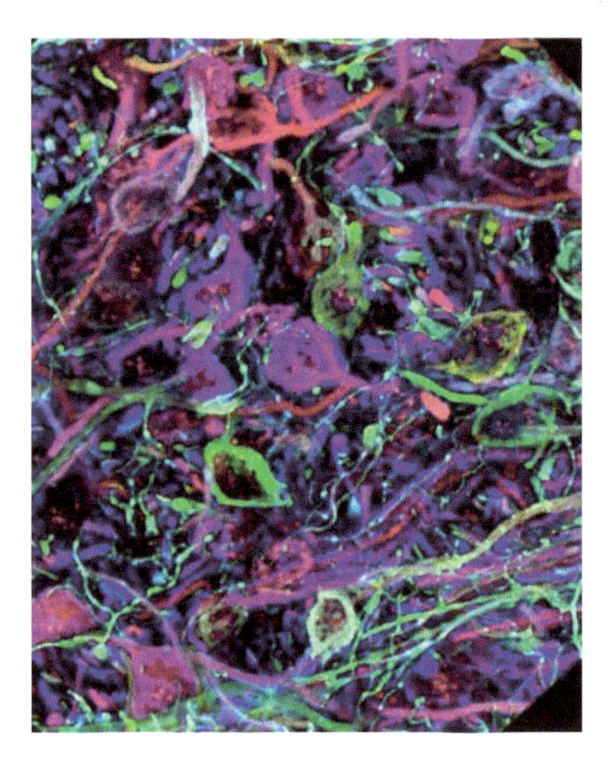

Figure 9.3 In this confocal microscopy image from the brainstem, the tube-like structures are the axons of the auditory pathway. Mammals have very thick axons in this region, which allows sound to be processed very quickly. (8) (Confocal image courtesy of Jean Livet.)

work with "dead" tissue, but he suggests that even this may change. "Scale is currently limited to fixed biological samples," he says, "but at some point in the future, there may be 'live Scale.'"

In 2013, two additional techniques to produce transparent brains were published, CLARITY and SeeDB. Neither of the methods is easy to use, and it takes weeks of practice to get a clear brain. The last step in the CLARITY process is the extraction of the lipids (fats) that make the brain opaque, a difficult step that requires the use of an electric field and negatively charged micelles. Kwanghun Chung, the Stanford postdoctoral student responsible for most of the CLARITY work, says this is the most challenging step: "We melted and burned hundreds of mouse brains before we figured out the optimum conditions, such as the voltage level." (11, p. 9) I can't imagine that the resultant smells made Kwanghun very popular among his labmates.

Scale, CLARITY, SeeDB, and Brainbow all require euthanization of the mouse and examination of its brain under a microscope. This is a severe limitation. Fortunately, there are many other imaging techniques that light up the brains of living organisms.

In the living brain, each behavior requires hundreds of thousands or even millions of neurons acting together in a synchronous fashion. Classical techniques only allow us to see what is happening to one neuron at a time, while with fluorescent proteins and optical techniques, we can monitor large numbers of neurons at the same time.

In 2005, the Howard Hughes Medical Institute's Janelia Farm Research campus was opened. It is a beautiful complex, with the windows of the main building serpentined along the contours of the Virginia hillside, seamlessly conforming to the site's topography. Inside the curving glass walls, the corridors are punctuated by soaring glass-enclosed staircases, and the roofs are covered with plantings. The 900-foot-long building, which would be the same height as an 85-story building if stood up, has three floors of research labs terraced into the gentle slope of the hillside. The aim of the billion-dollar facility and the 20 research groups housed therein is to discover the basic rules and mechanisms of the brain's information-processing system and to develop optical, biological, and computational technologies for creating and interpreting biological

Figure 9.4 Scale2 renders the mouse brain transparent without changing its shape or proportions, and without decreasing the intensity of fluorescence emitted by genetically encoded fluorescent proteins expressed in the tissue. (*10*)

images. Much of the research described in this chapter was performed at Janelia Farm.

According to Karel Svoboda, a group leader at Janelia Farm, paradigm shifts in neuroscience are strongly linked to methodological advances, and one of the biggest technological advances in the field has been the development of fluorescent protein technology. (12) He should know because he has been using fluorescent proteins to watch mice learn. His mice have glass windows replacing parts of their skulls and are genetically modified so the neurons that collect and process data from their whiskers are fluorescent green. The mice can live their entire lives with the windows in place, allowing Svoboda the opportunity to monitor any changes occurring over many weeks. He is particularly interested in the synapses, the tiny junctions between neurons, and has been observing them on a daily basis for months. (13a) "It's a very powerful technique that can look deep into the brain without disturbing it," Svoboda says. "There is something like a hundred billion synapses in the mouse brain and now we have some tricks to locate the same synapse each time we put the mouse under the microscope. It took a while to figure out, but now it's pretty routine." (13b) Being able to find the same neurons each time he looks at the mice has allowed him to observe that over time tiny spines along the dendrites are rising and receding. The older the mouse, the slower the spines grow. By trimming the whiskers on one side of the mouse, he is able to observe the rewiring of the brain associated with the mouse learning to use only one side of its whiskers. The rate of spine turnover increased as the mice were exposed to new experiences, such as being placed in a new maze. "The spines probably establish new synapses. If the synapse is a useful one, the spine will stay, if not it will retract. The neurons are constantly exploring alternative arrangements, which probably has something to do with learning," Svoboda said in an interview with *Discover* magazine. (14, p. 10) Svoboda did this work before he came to Janelia Farm; at the time he was still at the Cold Spring Harbor Laboratory on Long Island, New York. His fluorescent dendrites inspired the renowned glass artist Dale Chihuly to produce an 800-piece sculpture entitled *Twisting Dendrites* that weighs nearly half a ton. It hangs in the Marks Laboratory at Cold Spring Harbor, where its green glow greets the incoming neuroscientists (figure 9.5).

In Svoboda's experiments we see how tiny spines along the dendrites grow and recede as the mouse adapts to sensing its environment with a changing number of whiskers. When I was an undergraduate, we were taught that although dendrites could rearrange themselves, the adult brain could not grow new brain cells. In the last few decades, that has changed, and we now know that new brain cells can form in the

Figure 9.5 Dale Chihuly's *Twisting Dendrites*, a sculpture composed of 800 pieces of hand-blown glass, was inspired by the green fluorescent dendrites studied at the Cold Spring Harbor Laboratories. (Photo courtesy Miriam Chua, Cold Spring Harbor Laboratories.)

hippocampus, which is associated with memory and with the olfactory bulb that processes smells. In 2012, Seth Blackshaw and his students in the Johns Hopkins University Department of Neuroscience found signs indicating that nerve cells in the median eminence of the hypothalamus, which is involved with regulating sleep, hunger, and thirst, might also be forming in adults. They suspected that cells called *tanycytes* were rapidly proliferating and then differentiating into neurons, so they created some mice in which only the tanycytes expressed yellow fluorescent protein, and soon they had mice with fluorescent neurons. Because the hypothalamus has been implicated in obesity and metabolic disorders, Blackshaw's group compared neurogenesis in mice fed a high-fat diet with mice fed a regular diet and found that a high-fat diet will lead to adult mice producing four times as many neurons as regular mice (figure 9.6).

To establish the function of the new fluorescent neurons, a laser was used to kill off the newly produced neurons. The neurons were definitely up to no good; mice whose new neurons were neutralized were more active and didn't gain as much weight as the mice that continued to make

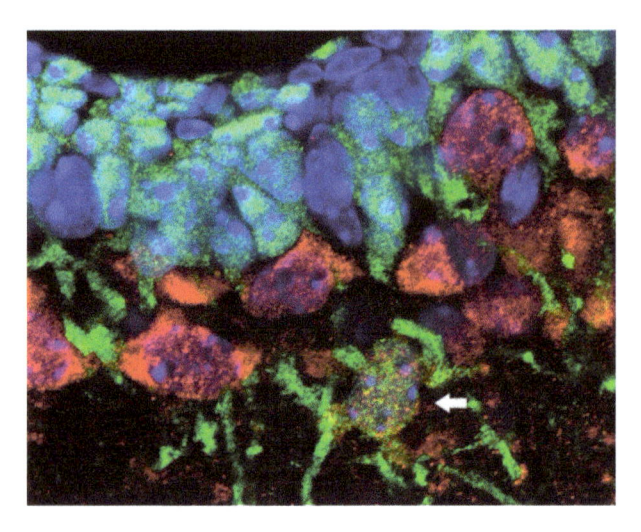

Figure 9.6 White arrow shows a new neuron that has developed from yellow fluorescent tanycytes. A fatty diet results in a fourfold increase of these new fat-loving brain cells. (15)

functioning brain cells in the median eminence. Blackshaw thinks the new neurons might be an evolutionary mechanism designed to make sure that the mice make the most of the rich, abundant, and in this case unfortunately unhealthy, food source they have found. He hopes that his findings may one day lead to a treatment for obesity in humans. (15)

Neurons are designed to transmit information; there are motor neurons that send signals from the brain to the muscles, sensory neurons that send information from sensory receptor cells all over the body to the brain, and interneurons that are responsible for transmissions between neurons. The signals they send can be both electrical and chemical. Although neurons vary in size and shape, they are all made up of four parts: axons, axon termini, dendrites, and cell bodies. As predicted by Cajal, the signals are unidirectional and go from the axon terminal of one cell to the dendrites of the next. And as we have seen in Svoboda's whisker experiments, some neurons have very few dendrites, but others have highly branched dendrites and transmit more information. Neurons differ in length; the longest in humans extends from the base of the spine to the big toe. Using Brainbow, we can track the paths of individual neurons in even the most densely tangled neurons, but Brainbow is not very good at imaging the connection of the axon terminal of one neuron to the dendrite of the next one in a neural synapses. That is where the technique known as mammalian GFP reconstitution across *synaptic partners* (*mGRASP*) comes in, as it is the technique of choice for mapping neurons and their synapses. Like

Brainbow, mGRASP makes use of the many colors found in fluorescent proteins.

In the first experiments that were done to show that the technique could be used to map mammalian synaptic connectivity in mice, Jinhyun Kim from the Center for Functional Connectomics at the Korean Institute of Science and Technology in Seoul, Korea, and Jeffrey Magee from the Janelia Farm Research Campus in Virginia and their co-workers used red, blue, and split green fluorescent proteins. On the presynaptic side of the synapse, the blue fluorescent protein, mCerulean, was tagged onto a transmembrane protein found only in the axon terminal of the neurons of interest, and a fragment of green fluorescent protein was attached to the end of the mCerulean. When excited with blue light, the presynaptic mGRASP component fluoresces blue; the GFP fragment doesn't fluoresce because it is incomplete and has no chromophore, the part of the protein that fluoresces and that cost Douglas Prasher his Nobel Prize. On the other side of the synapse, the postsynaptic side, a red fluorescent protein, dTomato, was attached to a transmembrane protein found only in dendrites, and the complementary GFP fragment to that found in the presynaptic component was fused to the end of dTomato. When excited, this fragment fluoresces red. In a typical human synaptic cleft, the axon and dendrites are separated by about 20 nanometers. By careful design, the two GFP fragments complement each other when the separation of the two mGRASP components is 20 nanometers, then they combine to form the complete, fully fluorescent GFP. In an mGRASP image one can therefore identify the presynaptic neuron by its blue fluorescence, the postsynaptic neuron by its red fluorescence, and the synapse by the green emission from the reconstituted split GFP (figures 9.7 and 9.8). (16) If

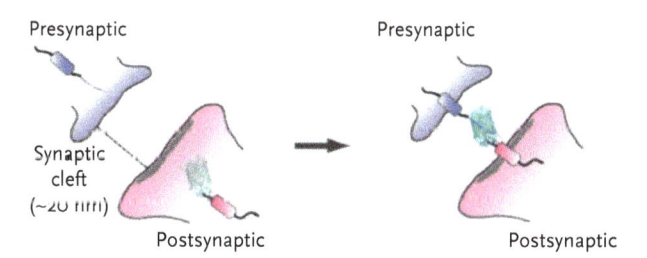

Figure 9.7 The presynaptic neuron has been genetically modified so that it has a blue fluorescent protein (the small blue barrel) and a small fragment of GFP extending into the synaptic cleft. The postsynaptic neuron has a red fluorescent protein (the small red barrel) and the missing part of a GFP molecule. When the two neurons are out of communication range, the two GFP fragments do not interact, and no green fluorescence is observed (left). If the two neurons are within chemical signaling range, the two GFP fragments combine to form the complete GFP molecule with its chromophore, and green fluorescence is observed (right). (16)

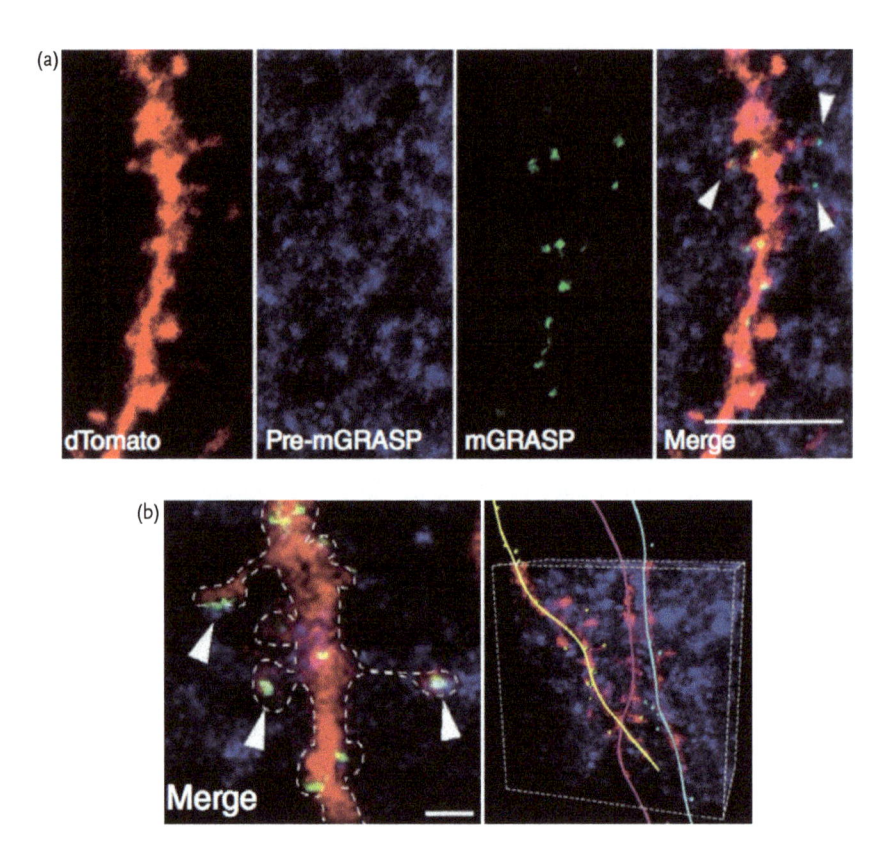

Figure 9.8 Imaging of mouse synapses in the hippocampus. The three leftmost images are photomicrographs of the same region of the brain taken with different filters to show separately the red dendrites (top far left), the blue axon terminals (top second from left), and the green synapses where the split GFP have joined (top second from right). The image on the far right is a superimposition of the three images. The arrowheads show where the blue axons intersect with the red dendrites to form the synapses that are expanded in the image shown in bottom left. The mGRASP technique can be extended to rapidly find the location of synapses in a three-dimensional brain sample (bottom right). (*16*)

there is no green fluorescence, then the blue axon and the red dendrite are not within 20 nanometers of each other.

The Janelia Farm and Korean Institute of Science and Technology researchers are currently working on a new improved mGRASP system that can be used in conjunction with the Brainbow and brain-clarifying techniques mentioned earlier. They have described their work as follows: "The method can allow rapid and precise characterization of synaptic connectivity in neuronal circuits in conditions of health as well as in models of neurological disorders that may be caused by abnormal synaptic connectivity, such as autism.. . . Our optimized mGRASP system, combined with computer-based three-dimensional (3D) reconstruction of neurons, will complement electron microscopy and optogenetic efforts toward an integrated 3D brain atlas, and can greatly accelerate comprehensive studies of synaptic long-range circuits and microcircuits." (*16*, p. 101) As of fall 2013, about 200 labs have requested the mGRASP virus.

Neurons get most of the attention in neuroscience, but in fact, they account for only 10 percent of the cells in the brain. Glial cells outnumber neurons by at least five to one. At first it was thought that glial cells did nothing but act as the glue of the nervous system, hence their name, which is Greek for "glue." That has changed as many new functions of glial cells have been found.

The blood-brain barrier is very useful because it protects the brain from toxins and pathogens, but it also prevents the immune system from entering the brain (that is where microglia come in; they are the brain's immune system). Alex Nimmerjahn, a biophysicist at the Salk Institute in La Jolla, California, has been recording the movements of these small glial cells. He says, "They're very dynamic, much more than any other cell in the adult brain." (*17*, p. 571) Nimmerjahn calculated that microglia can scan the whole brain every couple of hours, continually eating up invaders and damaged tissues and pruning away weak synapses. A number of neurodevelopmental disorders have been associated with faulty microglia. In promising but still controversial studies, a distinct behavioral improvement was observed after replacement of faulty microglia in mice with Rett syndrome, an autism spectrum disorder, and with trichotillomania, a psychiatric disorder characterized by compulsive hair pulling. (*17*)

Monica Vetter, the chair of the Department of Neurobiology and Anatomy at the University of Utah, is very interested in the role played by microglia in glaucoma, a neurodegenerative condition of the retina. Her group has shown that microglia are activated in early stages of the disease. "We are directly testing the role of microglia in neuronal decline, and defining the signals leading to their recruitment and activation with disease progression. Our ultimate goal is to identify key molecular pathways that can be targeted to slow or prevent blindness in glaucoma," she says. (*18*) Figure 9.9 shows some nerve fibers with their attendant GFP-labeled microglia as the fibers converge to form the optic nerve in a mouse retina. The photograph was one of the top 10 images of the first-ever BioArt competition sponsored by the Federation of American Societies for Experimental Biology. (*19*)

Astroglia cells are the most abundant cells in the human brain. These star-shaped glial cells are 20 times larger in humans than in mice, and human astroglia are also much more complex than their murine analogs.

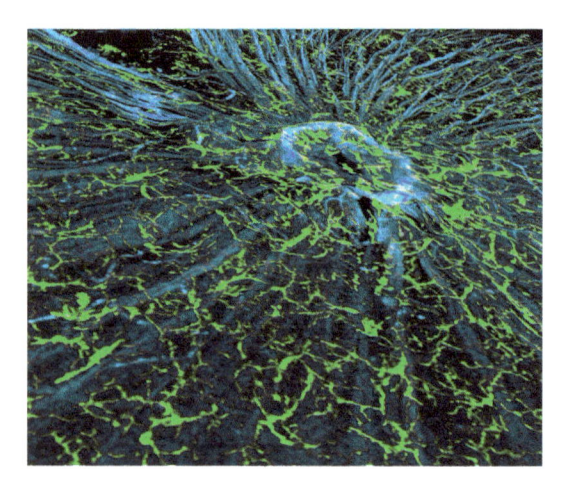

Figure 9.9 In early stages of glaucoma, microglia undergo structural changes that lead to irreversible neuronal decline. GFP-expressing microglia can be monitored as the fibers, which they are associated with, converge to form the retina. (19)

This observation has led some neuroscientists to speculate that human cognition is vastly superior to that of mice due to humans' advanced astroglial evolution. Maiken Nedergaard, a professor of translational neuromedicine at the University of Rochester, is one of them. In describing the impetus for her research, in which she implanted human GFP-expressing glial progenitor cells into newborn mouse brains, she said, "We were very interested in why astrocytes expanded so much during evolution. We were speculating that since they are larger, they may be integrating more information, making the human brain more powerful for computational analysis." (20) The experiments done with her University of Rochester colleague Steven Goldman and their students were a success. In fully grown mouse brains they observed that the implanted human glial progenitor cells had matured into green fluorescent astroglia that maintained their greater size and complexity, and that these mice were smarter than their wild-type brethren, outperforming mice grafted with murine glial progenitor cells in four standard long-term memory and learning tests. The researchers not only generated clever mice but also created a new model system of the human brain for the study of neurodegenerative diseases, stroke, and epilepsy. (21)

The imaging techniques discussed here have all been designed to produce a three-dimensional atlas of the brain that will lead to a better understanding of how the brain works and, perhaps equally important, to help find the solutions to problems in diseased brains. Derivatives of the same methods have also been used to understand the origins of brain diseases and disorders.

As we age, we become more vulnerable to neurodegenerative diseases. There is strong evidence that protein misfolding may cause some of these diseases, such as Alzheimer's, Parkinson's, Lou Gehrig's, and Huntington's disease.

Fluorescent proteins, like GFP, have a characteristic barrel shape; they fold into this three-dimensional structure because of their amino acid sequence. Occasionally something can go wrong, and despite GFP having the correct amino acid sequence, it doesn't fold into the barrel shape. It is misfolded and the chromophore doesn't form; subsequently, the GFP won't fluoresce. When protein misfolding occurs in the brain cells, the consequence are much more dire.

Scientists at the Department of Energy's Los Alamos National Laboratory have discovered a new method for rapidly determining whether proteins have folded correctly. It works by fusing the gene for a specially designed GFP to that of the protein being studied, relying on the fact that the GFP will not fold correctly if the protein it is bound to folds incorrectly. Therefore, misfolded proteins will be bound to misfolded nonfluorescing GFP, whereas correctly folded proteins will be labeled with fluorescent GFP. One of the advantages of this method is that it can tell you whether a protein is folded correctly even if you don't know the protein's function; you just need to know the location of its gene. According to Geoffrey Waldo, the developer of the method, "This assay will be particularly useful to medical researchers doing drug development and also to scientists working in the emerging field of proteomics—the study of the structures and functions of all the proteins encoded by the genome." (22, p. 694)

Shortly we will see how Michael Hecht from Princeton University's chemistry department has been using Waldo's GFP folding assay in his Alzheimer's research for the last 10 years. The disease he is studying has been around for at least a thousand years. The Greeks and Romans described individuals with symptoms that modern experts attribute to Alzheimer's disease. It is the most common cause of dementia among older people, affecting approximately 10 percent of people over the age of 65. It compromises memory and thinking skills, and eventually a person with Alzheimer's disease can no longer do even the simplest tasks. The costs for society are significant as patients with advanced stages of Alzheimer's disease require around-the-clock care and assistance.

In 1906, Dr. Alois Alzheimer performed an autopsy on a 55-year-old woman who had died from an unusual mental illness. He found many unexplained clumps and tangles of fibers in her brain. Today we know that the tangles are neurofibrillary tangles and that the clumps are amyloid plaques. The tangles and plaques that Dr. Alzheimer found are still two of the three main physiological features associated with the disease that bears his name; the other feature is the loss of connections between nerve cells. As yet we have no idea how Alzheimer's starts, but about 10 years before the occurrence of memory loss and cognitive decline, abnormal plaques and tangles appear throughout the brain, and neurons work less efficiently and eventually die. The tangles inhibit the transport of nutrients and key materials in the brain; then, as the neurons die, the brain shrinks, affecting nearly all its functions. The hippocampus, which is responsible for the creation of new memories, is particularly affected by the shrinkage. That is why patients in the early stages of the disease have difficulty recalling recent events. It is only later, as the disease progresses and brain shrinkage occurs, that long-term memory disappears, speech difficulties become evident, and aggressive behavior is manifested. Ultimately, individuals with Alzheimer's lose the ability to perform even the simplest tasks and become bedridden.

Alzheimer's disease can only be definitively diagnosed after death, with an autopsy that reveals a shrunken brain interspersed with amyloid plaques. The prevailing opinion is that the plaques cause the disease and that clearing the plaques pharmaceutically will undo the ravages of the disease.

The main components of the amyloid plaques are two protein fragments called Aβ40 and Aβ42 (made up of 40 and 42 amino acids, respectively). A number of experiments have been performed that highlight the importance of these two proteins in the initiation of Alzheimer's disease. It has been shown that Aβ42 is responsible for protein aggregation both in the test tube and in the brain. Furthermore, transgenic fruit flies and mice have been created that express the Aβ protein fragments, and they have been shown to be cognitively and behaviorally impaired.

Michael Hecht and his colleagues have tagged the Aβ42 protein fragment with the GFP folding assay. When Aβ42, which is associated with Alzheimer's disease, aggregates to form a plaque, the fluorescent proteins don't have room to fold properly and cannot fluorescence. Using this technique. Hecht's lab has created a screen for the Aβ42's aggregation properties. It is an Aβ42-green fluorescent protein fusion in *Escherichia coli*, in which the ability of the green fluorescent protein reporter to fold and fluoresce is inversely correlated with the aggregation propensity of the Aβ sequence. With the screen, the group has been able to establish which mutations in the 42-amino-acid fragment are responsible for increased and decreased propensity for amyloid plaque formation. Some of the mutations responsible for increased aggregation that were found in the lab have also been found in medical case studies. For example, a single mutation in the 22nd amino acid has been found in cases of familial early-onset Alzheimer's disease; it also leads to increased Aβ aggregation. (23)

In a search for new drugs that would fight the underlying cause of Alzheimer's disease, researchers in Hecht's lab have used the screen to test a library of 60,000 compounds for aggregation inhibitors. They were looking for potential drugs that would prevent amyloid plaque formation; these would be compounds that prevent aggregation and produce fluorescence. The compound that induced the most fluorescence has been tested on Alzheimer's animal models, with preliminary results that are promising. (24)

Since the first human embryonic stem cells were isolated in the late 1990s, researchers have been trying to convert them into replacement neurons for those damaged by neurodegenerative diseases. It is not a simple process to convert stem cells to specific neurons because there are hundreds of distinctly different neurons that establish the diversity that is required for the formation of neuronal circuits. Freda Miller, a senior scientist in the Developmental and Stem Cell Biology Program at the Hospital for Sick Children Research Institute in Toronto, and her colleagues have shown that a common diabetes drug, metformin, activates the formation of new brain cells from existing neural stem cells in the hippocampus. Because this is the same area of the brain that is affected in the early stages of Alzheimer's disease, Miller and her colleagues are hoping that the drug will stimulate the growth of brain cells to replace those incapacitated by the hippocampal shrinkage associated with Alzheimer's. In her research she used GFP to show that metformin activates brain cell production in mice (the drug nearly doubled the number of new neurons formed). Perhaps more important, classic behavioral tests on the mice treated with metformin showed that the new neurons were functional and that treated mice formed new memories significantly faster than those given a saline solution. (25) These results may suggest why a recent study by Michal Beeri at the Mount Sinai School of Medicine in New York and his colleagues found that insulin in combination with other diabetes medication is associated with less Alzheimer's neuropathology and may lead to a new treatment for early Alzheimer's. (26)

The second most common neurodegenerative disease is Parkinson's, with a prevalence of 1 percent in persons over 60 years of age. The disease first shows its presence through movement-related symptoms, such as shaking, difficulty walking, rigidity, and slowness of movement. As the disease progresses, cognitive and behavioral symptoms arise, with dementia commonly occurring in advanced stages of the illness. Parkinson's disease is named after Dr. James Parkinson, who published the article "An Essay on the Shaking Palsy" in 1817, in which he described the progression of the disease in six patients: the abnormal posture and gait, resting tremor, and diminished muscle strength. (27) In 1912, Fredric Lewy found microscopic particles in the midbrain of patients who have died from Parkinson's disease. Today the finding of these Lewy bodies in an autopsied brain is considered sufficient evidence that the individual suffered from Parkinson's disease. There are no lab tests to definitively identify the disease in living persons. In the Lewy bodies, the protein alpha-synuclein misfolds, sticks to other proteins, and creates aggregates that are toxic to the brain cells; they inhibit the formation and activity of the neurotransmitter dopamine, which is produced in the midbrain and is vital for mobility. Currently, most Parkinson's patients control their dopamine levels by taking the drug L-dopa.

Pamela McLean and Bradley Hyman, physicians at the MassGeneral Institute for Neurodegenerative Disease (MIND), have used a split GFP to examine this misfolding. They attached half a GFP molecule to one alpha-synuclein and the other half to a second alpha-synuclein molecule. When two separate alpha-synuclein proteins misfold and interact, the GFP halves come together, forming the complete characteristic GFP barrel and chromophore, which fluoresces. Using this technique, McLean and Hyman and their co-workers have found that chaperones can reduce the amount of alpha-synuclein misfolding. Chaperones are proteins found all over the body whose function is to aid proteins to fold into their correct shapes. They are also capable of refolding misfolded alpha-synuclein proteins. As McLean describes the action of the chaperones in the midbrain: "Imagine a chaperone at a dance separating two clingy teenagers, and you will have a vivid image of the brain's chaperone defense. We are researching whether drugs might activate additional chaperone proteins as a way to protect the brain against the ravages of alpha-synuclein and their resulting diseases." (28) Currently McLean and Hyman are using their alpha-synuclein-split GFPs to screen a series of drugs that are known to increase chaperone expression. (29)

Hyman and McLean and their co-workers have also created transgenic mice that express GFP-tagged human alpha-synuclein in their brains. In live mice they have been able to observe alpha-synuclein expression over a period of months; they have noted that different cortical neurons express varying amounts of the tagged protein and that in some neurons the alpha-synuclein is much less mobile than was previously thought. (30) The underlying reasons for these differences are presently unknown.

Another disease linked to misfolded proteins is early-onset dystonia, which is associated with repetitive movements, muscle contractions, and abnormal postures. Michael Sharp was struck with dystonia when he was 11 years old. His muscles refused to cooperate: they would lock up, and moving them was tremendously painful. He had little to no control over his body, which was twisted and contorted by the muscular contractions. Once his arm was locked in the same position for months. Michele Tagliati, a professor and the director of movement disorders at Cedars-Sinai Medical Center described Michael's condition when he first met him: "He could not sit on a chair, he had to lie down in a bed or on the floor. He could not write his name on a piece of paper." (*31*) There is no cure for dystonia, and treatment is limited to minimizing the symptoms. After Michael had suffered for 4 years and tried all known medications, his medical team decided to try deep brain stimulation, an invasive procedure in which electrodes are implanted in the brain, where they send electrical impulses to the part of the brain responsible for the muscular contractions. The changes have been remarkable, and Michael now attends law school. But the procedure doesn't work for everyone, and Michael is still fighting a daily battle. "The deep brain stimulation didn't get rid of dystonia completely. It is there, it's just being fought back against," he reported in an interview with KFSN-TV. (*31*)

Guy Caldwell and colleagues were studying early-onset dystonia when they stumbled on what they hoped might be a "molecular clog remover" that can remove misfolded proteins from brain cells and lead to a cure for early-onset dystonia. A mutated gene, TORA1, has been linked to early-onset dystonia. Caldwell and co-workers expressed the protein coded for by TORA1 in *Caenorhabditis elegans* to determine the function of the protein, which is called *torsin A*. When torsin A was expressed in worms containing aggregates of improperly folded proteins with GFP tags, the fluorescent clusters of misfolded proteins disappeared. Torsin A was acting like a roto-rooter, removing misfolded proteins. However, in *C. elegans* with the TORA1 mutation that is known to occur in early-onset dystonia, the mutated torsin A does not function properly, and the misfolded proteins are not removed. (*32*) Besides having found the cause of early-onset dystonia, Caldwell hopes that he might have stumbled across a protein that can remove aggregates of alpha-synuclein that are found in the brains of Parkinson's patients. Early results indicate that at least in *C. elegans* it works; torsin A chews up the malformed alpha-synuclein clusters too. However, *C. elegans* doesn't really have a brain, and as such it is not a very good model for human diseases.

Fortunately, Knud Larsen of the Department of Genetics and Biotechnology at the University of Aarhus, Denmark, has found the gene for torsin A in pigs. Not only is the pig brain an excellent model for the human brain, but Larsen and colleagues found that torsin A is very similar in both species, suggesting that the pig could be an ideal model for early-onset dystonia studies. (33)

In all neurodegenerative diseases, the disorder affects a very specific subset of neurons. For example, in Parkinson's disease, the mesencephalic dopaminergic neurons in the midbrain are destroyed, while all the other neurons surrounding it remain healthy and unaffected by the disease. Directing stem cells in a specific manner to produce a desired type of neuronal system has proved to be very difficult as the normal developmental pathways that generate most classes of central nervous system neurons are poorly defined and difficult to study. We currently know the most about the differentiation of spinal motor neurons and Parkinson's dopaminergic neurons from stem cells.

Vania Brocolli and his team of researchers at the San Raffaele Scientific Institute in Milan have managed to convert human and mouse skin cells into functional dopaminergic neurons. The neurons are functionally similar to primary dopaminergic neurons isolated from mouse brains. Furthermore, GFP-expressing skin cell–derived neurons become integrated and are functional when grafted into neonatal mouse brains. Unfortunately, the conversion rate for human skin cells is 10 to 20 times lower than that attained for mouse cells. This has led Brocolli and co-workers to focus on mouse cells. "Our goal is to use mouse cells first, then human cells in mice," he says. "If this works, we'll move to monkey models of the disease." (34, p. 227)

Spinal motor neurons are a central nervous system neuronal subtype for which pathways of neuronal specification have been defined. They are located in the rear of the spinal cord, from where their axons spread out and penetrate the muscles. The spinal motor neurons are the last step in the relay system that controls our posture and movements. Figure 9.10 shows a mouse embryo expressing green fluorescent protein in all its motor neurons. In patients with amyotrophic lateral sclerosis (ALS), also known as Lou Gehrig's disease, the motor neurons shrink and even disappear. As a result, the muscles can no longer be activated and begin to degenerate. There is great hope that in the near future we will be able to convert stem cells to motor neurons that can be used to treat ALS patients. This is significant because ALS affects about 25,000 Americans, with 5,000 new cases being diagnosed each year. The disease causes degeneration of the nerve cells in certain regions of the brain and spinal cord that control the voluntary muscles, which leads to loss of control of limb, mouth, and respiratory muscles. The median survival rate is 3 years, although some patients survive more than 10 years.

In large part this optimism is based on work by researchers at Columbia University who have engineered embryonic mouse stem cells that express GFP in the motor neurons. The modified stem cells were grown in vitro, and the scientists added two signaling proteins known to differentiate neural cells in live mice. First, retinoic acid was added to stimulate the formation of spinal cord cells from the stem cells; then a protein with the fantastic name sonic hedgehog was added to convert the spinal cells to spinal motor neurons. Sonic hedgehog is part of a family of proteins that direct cell development; for example, they determine how cells differentiate in the head and tail of a developing embryo. They were first discovered in fruit flies, where their absence resulted in embryos that were covered in spiny projections resembling the spines of hedgehogs. The first two mammalian proteins in the family were both named after hedgehog species, the desert hedgehog and Indian hedgehog; the third one to be found was named after the "Sonic the Hedgehog" video game. After the addition of sonic hedgehog, about 30 percent of the embryonic stem cells developed into motor neurons. But would they work like motor neurons? Hynek Wichterle, a postdoctoral researcher in Thomas Jessell's laboratory, tested the mouse motor neurons by inserting them into a chick's spinal neuron cord. They worked: green fluorescent mouse motor neurons grew long axons that connected with the intercostal muscles between the ribs (see figure 9.10). (35) In an interview in a Columbia University publication, Wichterle said, "I was pleasantly surprised at how well the stem cell–derived neurons

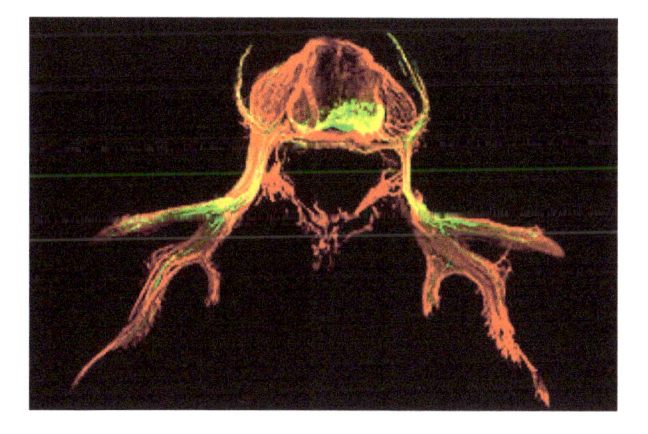

Figure 9.10 Chicken embryo expressing GFP in all its spinal motor neurons derived from mouse stem cells. (35)

mimicked the chick's neurons. But these experiments with embryonic stem cell–derived motor neurons are only the first step in our research for a potential treatment for amyotrophic lateral sclerosis. They open the way for subsequent experiments to determine which cells should be introduced into an adult animal with motor neuron degenerative disease." (36)

Unfortunately, there are some problems with embryonic stem cells, and it may be a while before they are used in human trials. The drawbacks associated with stem cells are their potential to form tumors, rejection by the host immune system, and regulatory and ethical concerns. This has led Wichterle and others to search for alternative sources of stem cells, such as the skin stem cells used by Vania Brocolli.

The vast majority of studies described in this book have used model organisms to study human diseases. But we all know that there are significant differences between mice and humans, and it isn't surprising that something that has worked well in mice, such as the conversion of skin cells to dopaminergic neurons, may not work well in humans. It would be useful to have a large number of diseased cells from a patient that could be studied and used in drug discovery. Hynek Wichterle has collaborated with Kevin Eggan of the Harvard Stem Cell Institute and the Department of Stem Cell and Regenerative Biology, and they have managed to take a skin cell from an 82-year-old patient diagnosed with a familial form of ALS and converted it to a pluripotent stem cell. Like embryonic stem cells, pluripotent stem cells are tremendously good at replicating and can be differentiated into any type of adult cell. Wichterle and Eggan took advantage of both these properties; they replicated their pluripotent stem cells and then exposed them to retinoic acid and sonic hedgehog so that all the pluripotent stem cells differentiated into motor neurons. In this way they could take a small number of skin cells from the patient, convert them to pluripotent stem cells, multiply them, and then change the large number of stem cells into motor neurons. The patient had a slowly progressing form of ALS with a rare known mutation occurring in its superoxide dismutase (SOD1) gene. The SOD1 mutation of the original donor was present in all the patient-derived neurons, thereby providing researchers with patient-specific disease neurons to study.

Because only 10 percent of all ALS cases are due to genetic mutations, the study of the other 90 percent of cases that are caused by unknown environmental and genetic factors is much more pressing. According to Kevin Eggan and his colleagues, their research described here will aid in the study "of living motor neurons generated from ALS cases with unknown genetic lesions, providing insight into their intrinsic survival properties, their interactions with other cell types, and their susceptibility to the environmental conditions that are considered to play an important role in ALS pathogenesis." (37, p. 1221)

This has just been a small sampling of the current research that uses fluorescent proteins to improve our understanding of neurodegenerative diseases and in turn searches for cures for these disorders. Every month, hundreds of papers are published that use bioluminescent proteins to add another piece to the big puzzles that need to be completed before we completely understand the causes and molecular consequences of important diseases like Alzheimer's. Another approach to studying neurodegenerative diseases is to examine the physical and mental changes that are associated with them. How does the brain of someone with ALS or Alzheimer's change as the disease progresses?

Gram for gram, the brain uses more energy than any other organ in the human body. This means that it needs a significant and constantly well-regulated flow of blood. The blood vessels at the base of the brain control the flow of blood to the brain. Since the brain receives approximately 20 percent of all the blood in circulation, these vessels control a fifth of the body's blood flow. Oxygenated blood is brought to active neurons, and deoxygenated blood is pumped to the lungs, where it is replenished. Functional magnetic resonance imaging (fMRI) relies on the magnetic differences between oxygenated and deoxygenated hemoglobin to see images of localized oxygen consumption in the brain associated with neural activity. This technique is particularly useful in establishing which areas of the brain are used for specific tasks, making it particularly suited to studies of neurodegeneration and its consequences. At present, fMRI can pinpoint and image areas of neural activity no smaller than 9 cubic centimeters, which is huge in comparison with the bioluminescent imaging methods discussed in this book. It is primarily a research tool to examine and locate areas of the brain associated with specific functions in healthy research subjects and is rarely used for biochemical research, although it is increasingly being used in medical diagnoses. The most commonly used fMRI approach involves measuring the blood oxygenation level dependency (BOLD), which is based on the fact that an increase in neuronal activity causes a locally decreased blood oxygenation, with an associated demand for more oxygen. The subsequent increase in blood flow and oxygenation to the neurons lowers the local deoxyhemoglobin concentrations, which leads to an increase of the fMRI signal. In BOLD fMRI, the difference in signal resulting from the oxygenation of the hemoglobin in the brain is measured. The oxygenated–deoxygenated–reoxygenated cycle associated with neural activity typically takes 20 to 30 seconds.

fMRI is the most commonly used noninvasive brain imaging technique. In this chapter I will highlight some applications of fMRI in research into ALS and Alzheimer's disease.

A series of fMRI studies have shown that larger volumes of the brain are active during finger tapping and flexing tasks of ALS patients than are observed in healthy control subjects. Similar results have been observed for many other physical tasks—it seems that the brain attempts to compensate, futilely, for the loss of neurons in the motor cortex caused by ALS disease by using new synapses and pathways. (38)

Detecting the onset of Alzheimer's disease as early as possible is very important to the treatment and understanding of the disease. Sterling Johnson of the Middleton Memorial Veterans Hospital in Madison, Wisconsin, and his coworkers have been using fMRI to show that functional brain changes may begin far in advance of symptomatic Alzheimer's disease. They used BOLD to monitor the blood supply to the brain of 74 healthy, middle-aged subjects with a family history and a genetic marker, which indicate that they were at risk of developing Alzheimer's disease. The fMRI experiments required the subjects to distinguish between previously viewed faces from a training set and new, previously unseen faces. The images were all grayscale photographs of forward-facing neutral faces (an equal number of male and female faces) taken from three different databases. During the experiment, participants lay on a scanner bed, wore protective earplugs and a high-resolution goggle system, and used a handheld two-button response box. Analysis of the resulting BOLD signals showed that both the genetic marker and the familial history were important predictors of fMRI patterns, and that they were relatively independent of each other in most brain regions (figure 9.11). The authors concluded, "Although many of these participants are at risk for Alzheimer's disease, only some will develop the disease, and such symptomatic decline may be years away. The clinical relevance of these studies will thus become clearer with longitudinal follow-up of changes in cognition and brain function. The purpose of the present report was to motivate this by pointing out baseline relationships between BOLD signal, cognition and Alzheimer's disease risk factors." (39, p. 390)

Every year we can see more of the brain. First it was Santiago Ramón y Cajal, who was able to trace the paths of a few neurons in slices of the brains of young birds. Then Brainbow came along and allowed its practitioners to track individual neurons in a tangle of brain cells based on their vibrant colors, while the green fluorescence of a recombined split GFP in mGRASP highlighted mouse synapses. But these are all structural

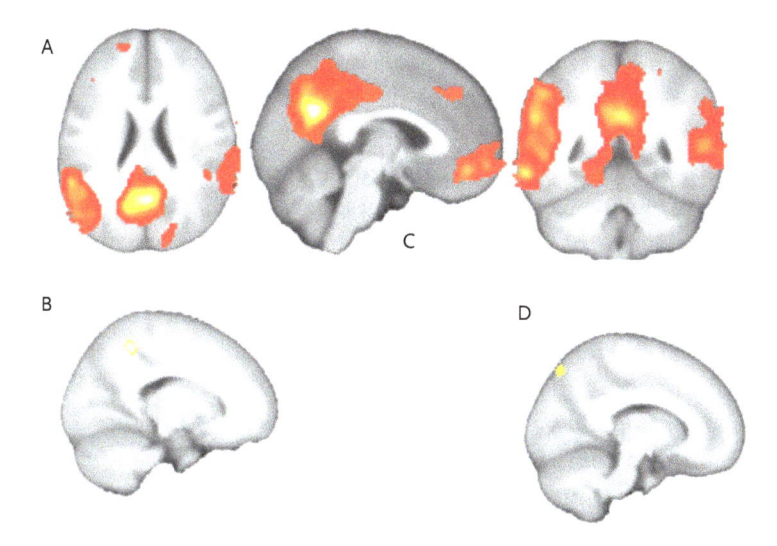

Figure 9.11 Previously viewed faces result in more neural activity than faces viewed for the first time in the three areas of the brain highlighted in A. The participants in the study without the ApoE gene that has been linked to Alzheimer's disease showed a higher signal in the region shown in B, while those with no familial Alzheimer's had a higher signal in the region shown in the image D. (*39*)

techniques—they allow us to see and follow the paths of individual neurons, but they don't tell us what the neurons are doing. fMRI can show us which areas of the brain are actively using oxygen; however, it does not have the greatest resolution. At best it can tell us about the activity in a cube of cranial matter no larger than 9 cubic centimeters. There is no way fMRI can be used to tell whether individual neurons are at rest or sending messages. When studying neurodegenerative diseases, this type of information would be extremely useful because knowing the function of individual neurons is critical in trying to understand how the brain works. Furthermore, visualizing when the ability of nerves to fire and send nerve impulses has been compromised will let us see the early signs of impending neurodegenerative disease. When a neuron fires, the concentration of the calcium ions changes at least a hundredfold; this is the largest concentration change of any of the molecules or ions in the neuron, which makes it the ideal indicator for neuronal activity.

Working with Atsushi Miyawaki and his colleagues, at the RIKEN Brain Science Institute in Japan, Roger Tsien, one of the three GFP researchers who won the Nobel Prize in Chemistry in 2008, has developed a cameleon that changes its color in the presence of calcium. (Unlike its namesake, this cameleon's name is spelled without an "h"—that's because the chemical symbol for calcium is Ca.) His cameleon does not have four legs, and it is not found in trees; instead, it is a molecule composed of two differently colored fluorescent proteins linked by a calcium-binding fragment of a protein called *calmodulin*, which changes its shape when it binds calcium ions. In the absence of calcium, the two fluorescent proteins of the cameleon construct are well separated and don't interact. Upon binding with calcium, however, the two fluorescent protein mutants are brought closer together, so that exciting one of the fluorescent proteins will influence the fluorescence of the other fluorescent protein. The more calcium that is present, the closer the two fluorescent molecules come to each other and the more they interact, causing a phenomenon called *fluorescence resonance energy transfer* (FRET). (40) These calcium sensors are called cameleons because they change color and have a long tongue (calmodulin) that retracts and extends in and out of the mouth when it binds and releases calcium.

In 2010 the groups of Atsushi Miyawaki, one of the developers of cameleon, and Mazahir Hasan (Department of Molecular Neurobiololgy, Max Planck Institute, Heidelberg, Germany) collaborated to virally infect the brains of mice with a new improved version of cameleon and they were able to measure the neural activity of entire regions of the brain during spontaneous and whisker-evoked action nerve impulses. (41) The setup shown in figure 9.12 illustrates how Hasan was able to record and locate all the neurons that are activated when the mouse whiskers are exposed to puffs of air. Thanks to the fluorescence of the cameleon construct, the authors were able to monitor neuronal activity in the brains of live mice. The experiment showed that cameleon reacted quickly and sensitively to fluctuations in calcium concentrations occurring in rapid sequence.

Obviously, blowing puffs of air on the whiskers of an immobilized mouse was only a proof of concept experiment. But it also was a steppingstone to

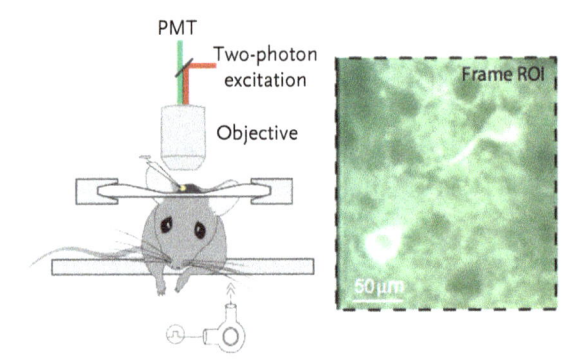

Figure 9.12 Setup for bulk recording of whisker evoked neural activity in mouse with cameleon construct. The image on the right shows two neurons that were activated with puffs of air in a 12-week-old mouse 6 weeks after it was virally infected with the cameleon calcium sensor. (41)

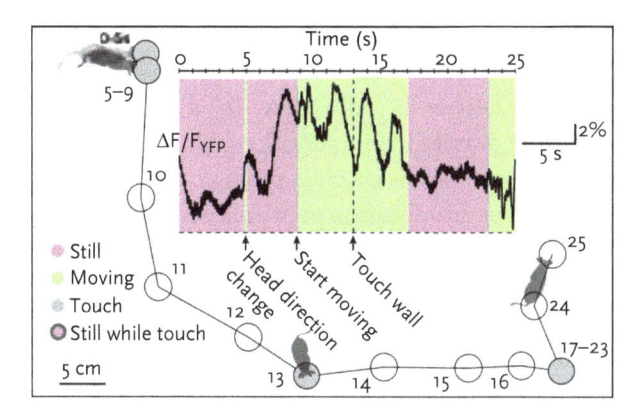

Figure 9.13 Fiber-optic recording of brain activity in freely moving mouse with cameleon calcium sensor. A graph of the change in fluorescence, which is related to the calcium concentration flux, versus time, is shown together with the position of the mouse in an open field box during the 25 seconds being graphed. The mouse behavior, sitting still, moving, touching, or having contact with the wall, is indicated by background colors. (*41*)

bigger, more important, and more complex experiments, such as trying to monitor spontaneous neural activity in a freely moving mouse. A similar setup was used for those experiments. The same researchers connected a single optic fiber to the head of an awake, freely moving mouse and measured the neural activity of individual neurons as the mouse freely moved around in an open field box (figure 9.13). A press release by the Max Planck Institute, where the majority of this research was conducted, concludes that "these new advances, using light to study the brain, provide us with a unique opportunity to investigate how memories are formed and lost and, furthermore, when and where nerve cell activity patterns become altered as in the case of aging and also in neurological diseases such as Alzheimer's disease, Parkinson's disease and schizophrenia." (*42*)

Karel Svoboda is a group leader at the Janelia Farm Research Campus. Something about him reminds me of a conductor; perhaps it is the way he uses his arms when he talks, or his Czech accent (he was born in Czechoslovakia and went to school in Germany). Or perhaps it's more cerebral, such as the fact that he likes to think of the object of his fascination, the brain, as an orchestra. According to Svoboda, imaging tools, such as Brainbow, that light up neurons are very useful, but they are like photographs of the instruments in the orchestra. He says, "To fully understand the orchestra you need to hear the individual instruments as they play their part in the symphony." (*12*)

When Loren Looger arrived at Janelia Farm, Svoboda asked him whether he could build a protein that would signal the presence of calcium

DISEASES OF THE BRAIN [193]

inside cells; Svoboda wanted an alternative to cameleon, something that would record the individual instruments. Looger doesn't look like a member of Svoboda's orchestra; he looks more like a member of a rock band. He is never still and dresses in vibrant colors. It's hard to believe that this boisterous multitasker was once a child prodigy. According to his mother, Looger started doing multiplication problems when he was 3 years old; as a young teenager he took part in mathematics and chemistry Olympiads before earning a BS degree in chemistry and an MS degree in mathematics from Stanford University in just 4 years. Svoboda calls him a samurai. "Loren is the consummate collaborator. He has a very unique skill set, and he is looking for damsels in distress," Svoboda says. "He's the kind of person who loves getting involved in other people's problems, in the very best sense." (43, p. 20)

It took Looger a few years to develop the calcium sensor that Svoboda was looking for. With this neuronal calcium sensor, called GCaMP3, it is possible to shine a light on the brain and look at blinking neurons with a microscope. One can see the music played by the whole neuronal orchestra, with some neurons lighting up in harmony as others fade away—a visual symphony of our thoughts. And by analyzing the musical "score," one can determine how the neurons play together and discover what happens when some of them are deficient. Svoboda calls this monitoring of neural activity the second wave of the GFP revolution.

Looger has been collaborating with many researchers to show the utility of his GFP-based sensor in neural imaging. Vivek Jayaraman and co-workers at Janelia Farm have used the sensor to track neural activity in the olfactory system of the fruit fly brain; Cornelia Bargmann's lab has watched neurons in *C. elegans*, a flatworm, light up as certain smells were presented and taken away from it; and Svoboda's own group has observed 13 mouse neurons that lit up in a particular sequence as the mouse walked and moved its whiskers. (44) A video of the neurons lighting up as they fire in sequence when the mouse moves a single whisker is reminiscent of the night sky during a thunderstorm, with flashing neurons illuminating the brain. The video (which can be seen at www.hhmi.org/news/looger20091108.html) reveals that the neurons being imaged seem to be performing a wide range of tasks.

For Looger, this is just the beginning. He believes that "for the foreseeable future we need to focus on simple-ish, but profoundly important questions. When the mouse looked left, which thousand neurons were used? Which ones came on first? Then which ones next? Was it repeatable, so when the mouse did it again, was it the *exact same* thousand neurons? If two animals are under the microscope, do they both use the same thousand neurons to do this?" (45) Looger wants to tease out the contributions of individual neurons to a simple behavior, which for him is the first step in cracking the functional wiring diagram of the brain.

Teaching by repetition and practice may not be the most modern of pedagogical techniques, but it works. Karel Svoboda has shown that training mice to do a new task causes neurons to work together more efficiently. Together with his postdoctoral researcher Takaki Komiyama, Svobada taught his mice to lick in response to one odor, but not when exposed to another, all the while reading their minds through a glass window in their skulls. The calcium sensor GCaMP3 lit up in two distinct areas of the mouse motor cortex, indicating that there are two groups of lick-controlling neurons. One set of neurons is probably in charge of the coordinated jaw and mouth movements, while the other group might directly control the tongue muscles. As the mice learn, the number of lick-controlling neurons in each group seems to decrease. (46) "We know that as we train to perform a behavior, it tends to become easier," Komiyama said. "Using fewer neurons more efficiently might be one way to explain that phenomenon." (47)

Short-term memories that are used in making decisions are critical factors in learning, reasoning, and comprehending materials; that is why they are often known as the *working memory*. Some neurological diseases such as schizophrenia may involve a malfunctioning working memory. David Tank and his colleagues at Princeton University's Bezos Center for Neural Circuit Dynamics have used Laren Looger's calcium sensor to study working memory in mice. Their experiments have changed the way we view short-term memory deposition and retrieval. The old belief that that entire populations of neurons involved in the working memory fire in unison has been replaced with the notion that sequential firing is critical to memory storage—the wrong sequence will result in false memory storage and recall.

Tank's mice were trained to walk or run through a virtual maze while on a spherical treadmill with their heads kept stationary, which is ideal for brain imaging. The virtual maze was displayed on a wide-angle screen surrounding the mouse. The maze was simple and consisted of a long passage that ended with a T-junction. As the mouse ran along passage, it was given a cue to turn left or right at the T-junction. If the mouse read the cue correctly and made the correct turn, it was rewarded with some water. After a few training runs the mice made the correct turn about 90 percent of the time. During the 10-second period it took the mice to form a memory of the cue, store it, and then act on it, a sequential neuronal firing pattern was observed in the posterior parietal cortex. A different sequence of neurons was activated for a left turn than for a right turn. By monitoring the calcium sensor fluorescence, Tank and his students were able to see when the mouse registered the cue and stored the memory. They could also see when the mouse stored the memory but then lost it and began firing the sequence for the wrong turn. The researchers were not surprised to discover that the posterior parietal cortex was involved because this part of the brain has been shown to be important in movement planning, decision-making, and spatial attention in humans and monkeys. (48) However, no one had expected that populations of neurons fire in distinctive sequences when the brain is storing a short-term memory. Prior to these experiments, the existing belief was that entire populations of neurons involved in the working memory fire in unison.

As is often the case in studies like this one, it is easy to be amazed by futuristic aspects of the experiments and to forget the big picture. "Studies such as this one are aimed at understanding the basic principles of neural activity during working memory in the normal brain. However, the work may in the future assist researchers in understanding how activity might be altered in brain disorders that involve deficits in working memory," Tank reminds us. (49)

Looking for a unique way to display your love to that special person in your life? Aravinthan Samuel and his group of researchers at the Harvard University Department of Physics and Brain Science may just have the solution to your quest; they have managed to direct *C. elegans* to wiggle around forming perfect hearts. Samuel's research is not meant to take a bite out of Hallmark's Valentine card market; rather, it is an attempt to understand how the nervous system converts sensory perception into movement. Humans have more than a billion neurons, which makes understanding what is going on in our brains rather complicated. In contrast, *C. elegans* has only 302 neurons, and we know where they are and where they go, making it an ideal research subject for studies in which even fruit flies are to complex.

C. elegans colonies are grown in petri dishes containing agar. Within seconds of turning on an electric field, nearly every worm on the agar surface will make its way to the negative pole. This has been known since 1978, but it was not until 2007 that researchers discovered that the 1-millimeter worms don't head straight to the negative pole; instead, they crawl in an undulating fashion following the force lines of the electric field, and the direction of their movement can be changed by altering the field. While a steady rotation will cause the poor nematodes to snake around in perfect circles, a more skillful electric field operator is be able to control the

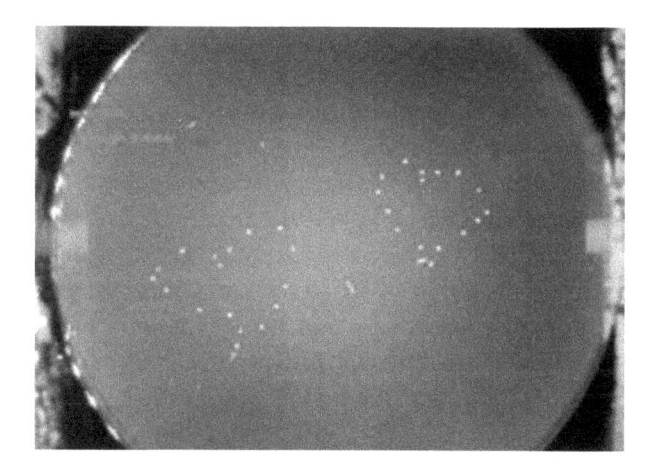

Figure 9.14 *C. elegans* moving along the force lines of an electric field. The nematodes have been genetically modified to express cameleon, an in vivo calcium sensor, in order to find the neurons that respond to the electric field. (Courtesy of A. Samuel.)

worms like lines on an Etch A Sketch (figure 9.14 and www.conncoll.edu/ccacad/zimmer/GFP-ww/cooluses11.html).

Samuel labeled specific sensory neurons with cameleon, the Tsien GFP protein construct that measures the calcium concentration in cells. In this way he was able to establish which sensory neurons are sensitive to the direction of the electric field, and which interneuronal circuits are responsible for the decision to use either a turn or a reversal to reorient the worm in response to a change in the electric field. This study is a significant step in understanding how the nervous system transfers sensory input into motor output. (50)

Roundworm, fruit fly, zebrafish, mouse, and monkey brains are used as model systems for what goes on in the human brain. Each organism has its supporters. Recently, zebrafish have been receiving a lot of attention because they are transparent, they record and store memories just like mammals do, and, unlike in mammals, their neural tubes fold outward, resulting in important areas of the brain being located on the surface. Furthermore, current imaging techniques have just progressed to the point where they can produce complete brain-activity maps of the zebrafish brain. In 2013, the laboratories of Janelia Farm microscopist Phillipp Keller and neurobiologist Misha Ahrens reported that they had developed a system that is capable of recording the activity of more than 80 percent of the fish's 100,000 neurons. The activity of all the neurons can't be monitored because some neurons are located between the eyes, where it is hard to image neural activity at the cellular level. Unlike a conventional microscope that uses a beam of light, the new technique use sheets of light and a detector that measures the neural activity every 1.3 seconds. (51) Light-sheet microscopy allows researchers to image at least 50 times more neurons than conventional microscopy. "We see the big picture without losing resolution," says Keller. (52)

Twenty years after Martin Chalfie's fluorescent *C. elegans* first graced the cover of *Science*, the field of fluorescent proteins is still growing. Every month, new techniques based on fluorescent proteins are published, new cell lines are released, and new microscopes and filters enter the market. It is difficult not to be amazed by miniaturized fluorescent microscopes weighing less than 2 grams that are designed to sit on top of the heads of live mice and two photon imaging setups that can record changes in the calcium concentration as fruit flies walk or fly around. (53)

In 1885–1886, Santiago Ramón y Cajal wrote 12 short stories that explored ethical aspects of new developments in science—bacteriology, artificial insemination, and the power of suggestion. In 1905, he published five of these stories in a book entitled *Vacation Stories*. Due to the book's sexual content, diabolical humor, and unflattering caricatures of priests, he published it under the pseudonym "Dr. Bacteria," and to protect his reputation, he did not publish the other seven stories, not even under his alias.

Laura Otis of Emory University has translated the five short stories in *Vacation Stories* into English. Cajal's first story, "For a Secret Defense, a Secret Revenge," describes the actions of an aging bacteriologist who suspects his young wife is having an affair with his lab assistant. He sets up a seismographic device to measure the vibrations of the laboratory couch, and his worst fears are confirmed; it seems the couch has witnessed some seismic activity. The bacteriologist carefully plans his revenge. He infects the assistant with a bovine tuberculosis bacillus, waits to see whether it will be transmitted to his wife, and, when it does, writes up the results in a bacteriological journal. (54)

This is quite a story, especially if you consider that it was written in 1885! It would be very interesting to know what "Dr. Bacteria" would write now, if he knew about modern science with its fluorescent calcium sensors and fMRIs that can light up the mind, making mind-reading more science than fiction. I am sure he would have been charmed with the idea that his concepts were the foundations of the present-day mind-reading experiments described in this chapter and the mind control experiments described in the next chapter. It is hard not to be amazed with all the research described in this chapter, but at the same time it is daunting to think about how much there is still to know and how little we really know. Some things haven't changed much since Cajal's times. As V. S. Ramachandran, a leading neuroscientist, says,

"The brain is a 1.5 kilogram mass of jelly, the consistency of tofu, you can hold it in the palm of your hand, yet it can contemplate the vastness of space and time, the meaning of infinity and the meaning of existence. It can ask questions about who am I, where do I come from, questions about love and beauty, aesthetics, and art, and all these questions arising from this lump of jelly. It is truly the greatest of mysteries." (55, p. 15)

REFERENCES

1. Everdell, W. R. (1997). *The first moderns: Profiles in the origins of twentieth-century thought*. Chicago: University of Chicago Press.
2. Mann, G. (1902). *Physiological histology, methods and theory*. Oxford: Clarendon Press.
3. Ramón y Cajal, S., Craigie, E. H., and Cano, J. (1937). *Recollections of my life*. Philadelphia: American Philosophical Society.
4. Llinas, R. R. (2003). The contribution of Santiago Ramón y Cajal to functional neuroscience. *Nature Reviews Neuroscience 4*, 77–80.
5. The Nobel Prize in Physiology or Medicine 1906 (2014). http://www.nobelprize.org/nobel_prizes/medicine/laureates/1906/.
6. Nobelstiftelsen. (1964). *Physiology or medicine*. Amsterdam: Published for the Nobel Foundation by Elsevier.
7. Bradt, S. (2007). Scientists image vivid 'brainbows', Harvard gazette, http://news.harvard.edu/gazette/story/2007/11/scientists-image-vivid-'brainbows'/.
8. Livet, J., Weissman, T. A., Kang, H. N., Draft, R. W., Lu, J., Bennis, R. A., Sanes, J. R., and Lichtman, J. W. (2007). Transgenic strategies for combinatorial expression of fluorescent proteins in the nervous system. *Nature 450*, 56–62.
9. Hopkin, K. (2011). Critical connections. *The Scientist 25*, 54–56.
10. Hama, H., Kurokawa, H., Kawano, H., Ando, R., Shimogori, T., Noda, H., Fukami, K., Sakaue-Sawano, A., and Miyawaki, A. (2011). Scale: A chemical approach for fluorescence imaging and reconstruction of transparent mouse brain. *Nature Neuroscience 14*, 1481–1488.
11. Wolf, L. K. (2013). Chemical method that makes tissue transparent could lead to a brain wiring diagram. *Chemical and Engineering News* April 15, *91*, 9.
12. Svoboda, K. (2012). Personal communication.
13a. Trachtenberg, J. T., Chen, B. E., Knott, G. W., Feng, G. P., Sanes, J. R., Welker, E., and Svoboda, K. (2002). Long-term in vivo imaging of experience-dependent synaptic plasticity in adult cortex. *Nature 420*, 788–794.
13b. Svoboda, K. (2013). http://www.hhmi.org/scientists/karel-svoboda.
14. Svitil, K. (2003). Memory's machine. *Discover 24*, 10.
15. Lee, D. A., Bedont, J. L., Pak, T., Wang, H., Song, J., Miranda-Angulo, A., Takiar, V., Charubhumi, V., Balordi, F., Takebayashi, H., Aja, S., Ford, E., Fishell, G., and Blackshaw, S. (2012). Tanycytes of the hypothalamic median eminence form a diet-responsive neurogenic niche. *Nature Neuroscience 15*, 700–702.
16. Kim, J., Zhao, T., Petralia, R. S., Yu, Y., Peng, H. C., Myers, E., and Magee, J. C. (2012). mGRASP enables mapping mammalian synaptic connectivity with light microscopy. *Nature Methods 9*, 96–102.

17. Hughes, V. (2012). The constant gardeners. *Nature 485*, 570–573.
18. http://neuroscience.med.utah.edu/Faculty/Vetter.html
19. Bosco, A., Steele, M. R., and Vetter, M. L. (2011). Early microglia activation in a mouse model of chronic glaucoma. *Journal of Comparative Neurology 519*, 599–620.
20. Gorman, M. G. (2013). Mice get smarter with human brain cells. *BioTechniques*, http://www.biotechniques.com/news/Mice-Get-Smarter-with-Human-Brain-Cells/biotechniques-340903.html?service=print#.U4eM9VzLOv0.
21. Han, X., Chen, M., Wang, F., Windrem, M., Wang, S., Shanz, S., Xu, Q., Oberheim, N. A., Bekar, L., Betstadt, S., Silva, A. J., Takano, T., Goldman, S. A., and Nedergaard, M. (2013). Forebrain engraftment by human glial progenitor cells enhances synaptic plasticity and learning in adult mice. *Cell Stem Cell 12*, 342–353.
22. Waldo, G. S., Standish, B. M., Berendzen, J., and Terwilliger, T. C. (1999). Rapid protein-folding assay using green fluorescent protein. *Nature Biotechnology 17*, 691–695.
23. Kim, W., and Hecht, M. H. (2008). Mutations enhance the aggregation propensity of the Alzheimer's Aβ peptide. *Journal of Molecular Biology 377*, 565–574.
24. Kim, W., Kim, Y., Min, J., Kim, D. J., Chang, Y. T., and Hecht, M. H. (2006). A high-throughput screen for compounds that inhibit aggregation of the Alzheimer's peptide. *ACS Chemical Biology 1*, 461–469.
25. Wang, J., Gallagher, D., DeVito, L. M., Cancino, G. I., Tsui, D., He, L., Keller, G. M., Frankland, P. W., Kaplan, D. R., and Miller, F. D. (2012). Metformin activates an atypical PKC-CBP pathway to promote neurogenesis and enhance spatial memory formation. *Cell Stem Cell 11*, 23–35.
26. Beeri, M. S., Schmeidler, J., Silverman, J. M., Gandy, S., Wysocki, M., Hannigan, C. M., Purohit, D. P., Lesser, G., Grossman, H. T., and Haroutunian, V. (2008). Insulin in combination with other diabetes medication is associated with less Alzheimer neuropathology. *Neurology 71*, 750–757.
27. Lees, A. J. (2007). Unresolved issues relating to the shaking palsy on the celebration of James Parkinson's 250th birthday. *Movement Disorders 22*, S327–S334.
28. Massachusetts General Hospital, Newsarticle (2009). Lighting up Parkinson's disease research: How jellyfish could potentially play a role in treatment. http://www.massgeneral.org/psychiatry/news/newsarticle.aspx?id=1627.
29. Kalia, S. K., Kalia, L. V., and McLean, P. J. (2010). Molecular chaperones as rational drug targets for Parkinson's disease therapeutics. *CNS & Neurological Disorders-Drug Targets 9*, 741–753.
30. Unni, V. K., Weissman, T. A., Rockenstein, E., Masliah, E., McLean, P. J., and Hyman, B. T. (2010). In vivo imaging of alpha-synuclein in mouse cortex demonstrates stable expression and differential subcellular compartment mobility. *PLOS One 5*(5), e10589.
31. Kim, M. (2012). Michael's miracle: Making dystonia disappear. *Health Watch*. http://abclocal.go.com/kfsn/story?section=news/health/health_watch&id=8590597. July 9.
32. Caldwell, G. A., Cao, S., Sexton, E. G., Gelwix, C. C., Bevel, J. P., and Caldwell, K. A. (2003). Suppression of polyglutamine-induced protein aggregation in Caenorhabditis elegans by torsin proteins. *Human Molecular Genetics 12*, 307–319.

33. Henriksen, C., Madsen, L. B., Bendixen, C., and Larsen, K. (2009). Characterization of the porcine TOR1A gene: The first step towards generation of a pig model for dystonia. *Gene 430*, 105–115.
34. Caiazzo, M., Dell'Anno, M. T., Dvoretskova, E., Lazarevic, D., Taverna, S., Leo, D., Sotnikova, T. D., Menegon, A., Roncaglia, P., Colciago, G., Russo, G., Carninci, P., Pezzoli, G., Gainetdinov, R. R., Gustincich, S., Dityatev, A., and Broccoli, V. (2011). Direct generation of functional dopaminergic neurons from mouse and human fibroblasts. *Nature 476*, 224–227.
35. Wichterle, H., Lieberam, I., Porter, J. A., and Jessell, T. M. (2002). Directed differentiation of embryonic stem cells into motor neurons. *Cell 110*, 385–397.
36. Conova S. (2002). Custom-made motor neurons. *In vivo Columbia University Health Sciences*, http://www.cumc.columbia.edu/publications/in-vivo/Vol1_Iss15_sept25_02/.
37. Dimos, J. T., Rodolfa, K. T., Niakan, K. K., Weisenthal, L. M., Mitsumoto, H., Chung, W., Croft, G. F., Saphier, G., Leibel, R., Goland, R., Wichterle, H., Henderson, C. E., and Eggan, K. (2008). Induced pluripotent stem cells generated from patients with ALS can be differentiated into motor neurons. *Science 321*, 1218–1221.
38. Lule, D., Ludolph, A. C., and Kassubek, J. (2009). MRI-based functional neuroimaging in ALS: An update. *Amyotrophic Lateral Sclerosis 10*, 258–268.
39. Xu, G. F., McLaren, D. G., Ries, M. L., Fitzgerald, M. E., Bendlin, B. B., Rowley, H. A., Sager, M. A., Atwood, C., Asthana, S., and Johnson, S. C. (2009). The influence of parental history of Alzheimer's disease and apolipoprotein E 4 on the BOLD signal during recognition memory. *Brain 132*, 383–391.
40. Miyawaki, A., Llopis, J., Heim, R., McCaffery, J. M., Adams, J. A., Ikura, M., and Tsien, R. Y. (1997). Fluorescent indicators for Ca2+ based on green fluorescent proteins and calmodulin. *Nature 388*, 882–887.
41. Lutcke, H., Murayama, M., Hahn, T., Margolis, D. J., Astori, S., Borgloh, S. M. Z., Gobel, W., Yang, Y., Tang, W. N., Kugler, S., Sprengel, R., Nagai, T., Miyawaki, A., Larkum, M. E., Helmchen, F., and Hasan, M. T. (2010). Optical recording of neuronal activity with a genetically-encoded calcium indicator in anesthetized and freely moving mice. *Frontiers in Neural Circuits 4*, 1–12.
42. Max Planck Gesellschaft. (2010). Blinking neurons give thoughts away. www.mpg.de/624454/pressRelease20100504.
43. Michalowski, J. (2012). Enter the samurai. *HHMI Bulletin*, May 19–23.
44. Tian, L., Hires, S. A., Mao, T., Huber, D., Chiappe, M. E., Chalasani, S. H., Petreanu, L., Akerboom, J., McKinney, S. A., Schreiter, E. R., Bargmann, C. I., Jayaraman, V., Svoboda, K., and Looger, L. L. (2009). Imaging neural activity in worms, flies and mice with improved GCaMP calcium indicators. *Nature Methods 6*, 875–881.
45. HHMI News. (2009). Teasing apart brain function, neuron by neuron. www.hhmi.org/news/looger20091108.html. May 29.
46. Komiyama, T., Sato, T. R., O'Connor, D. H., Zhang, Y. X., Huber, D., Hooks, B. M., Gabitto, M., and Svoboda, K. (2010). Learning-related fine-scale specificity imaged in motor cortex circuits of behaving mice. *Nature 464*, 1182–1186.
47. *HHMI News* (2010). Learning neurons fire together. http://www.hhmi.org/news/learning-neurons-fire-together.
48. Harvey, C. D., Coen, P., and Tank, D. W. (2012). Choice-specific sequences in parietal cortex during a virtual-navigation decision task. *Nature 484*, 62–68.

49. Zandonella, C. (2012). Princeton scientists identify neural activity sequences that help form memory, decision-making. www.princeton.edu/main/news/archive/S33/17/36M20/index.xml?section=topstories. May 29.
50. Gabel, C. V., Gabel, H., Pavlichin, D., Kao, A., Clark, D. A., and Samuel, A. D. T. (2007). Neural circuits mediate electrosensory behavior in Caenorhabditis elegans. *Journal of Neuroscience 27*, 7586–7596.
51. Ahrens, M. B., Orger, M. B., Robson, D. N., Li, J. M., and Keller, P. J. (2013). Whole-brain functional imaging at cellular resolution using light-sheet microscopy. *Nature Methods 10*, 413–420.
52. Baker, M. (2013). Flashing fish brains filmed in action. *Nature News*, http://www.nature.com/news/flashing-fish-brains-filmed-in-action-1.12621.
53. Marx, V. (2012). Rendering the brain-behavior link visible. *Nature Methods 9*, 953–958.
54. Ramón y Cajal, S., and Otis, L. (2001). *Vacation stories: Five science fiction tales.* Urbana: University of Illinois Press.
55. Ramachandran, V. S. (2012). Speaking of science. *The Scientist 26*, 15.

CHAPTER 10

Optogenetics

I control the brain in order to understand how it works... An experienced code-breaker will tell you that in order to figure out what the symbols in a code mean, it's essential to be able to play with them, to rearrange them at will. So in this situation too, to decode the information contained in patterns like this, watching alone won't do. We need to rearrange the pattern. In other words, instead of recording the activity of neurons, we need to control it. It's not essential that we can control the activity of all neurons in the brain, just some. The more targeted our interventions, the better.

Gero Miesenböck, *Re-engineering the Brain*, filmed July 2010,
posted November 2010, TEDGlobal 2010

Taken to its extremes, the culmination of the previous chapter can be considered a form of mind-reading in which modified fluorescent proteins are used to light up active brain cells. In science fiction the logical extension of this work would be to use the knowledge gained from probes like GCaMP to activate individual neurons and thereby control the mind. Until recently, the closest neuroscientists could come to this fictive goal was to stimulate brain cells with electrodes, but even with the finest electrodes they could never activate single neurons. Now, thanks to optogenetics, scientists can use light and an algae protein to turn individual neurons on and off instantly. Of course, these optogeneticists are not evil villains bent on controlling the minds of others. They have no nefarious master plans; they are more interested in using optogenetics and other imaging techniques to help us understand how the human mind actually works. They want to use optogenetic methods to map out the function of individual brain cells and manipulate defective neurons.

The first chapter of this book introduced you to Osamu Shimomura and his obsession with bioluminescence. Shimomura spent 40 years unraveling the photochemistry of the crystal jellyfish. His research was a perfect example of how the quest for basic knowledge can lead to very practical and useful techniques. Without Shimomura's work, most of the imaging experiments described in this book would not be possible. However, it took more than 25 years for green fluorescent protein to make the trip from the jellyfish to *Caenorhabditis elegans* and the cover of *Science* magazine.

Although there are many similarities in the development of green fluorescent protein imaging technology and that of optogenetics, there are also important differences, as described in this chapter. From the very beginning, Peter Hegemann, Gero Miesenböck, Karl Deisseroth, and Ed Boyden, the pioneers of optogenetics, set out to find a way to stimulate individual neurons or specific groups of neurons. Their quest to establish a technique that would allow them to understand the function of individual neurons was an excellent example of applied science. Yet it too relied on at least 150 years of basic research on algae's response to light, which ultimately resulted in Georg Nagel, Peter Hegemann, and Ernst Bamberg discovering the on switch the optogeneticists required.

In the late 1990s, while at Yale University, Gero Miesenböck started looking for a way in which he could use light to switch on individual neurons. Four years after he came up with the idea, headless fruit flies trying to escape a nonexistent danger proved that light could be used to excite specific neurons in a living, albeit headless, organism. Susana Lima, a graduate student in Miesenböck's laboratory, did the breakthrough experiments. She created a genetically modified fruit fly that expressed optically gated ion channels in just 2 of its 200,000 brain cells—not just any 2 neurons, but the neurons that are responsible for the flight response in fruit flies. It is these neurons that cause the fruit flies to fly away as soon as you move your hand into position to swat them. Lima's fruit flies behaved perfectly normally except that they flew away as soon as a flash of light excited the photo-activated rhodopsin and thus the brain cells in charge of the flight response. In a TED talk, Miesenböck, now at Oxford University, described how Lima controlled for the fact that the fruit flies might be seeing the flash of light and reacting to that: "Susana did a simple but brutally effective experiment. She cut the heads off of her flies. These headless bodies can live for about a day, but they don't do much. They just stand around and groom excessively. So it seems that the only trait that survives decapitation is vanity. Anyway, as you'll see in a moment, Susana was able to turn on the flight motor of what's the equivalent of the spinal cord of these flies and get some of the headless bodies to actually take off and fly away. They didn't get very far, obviously." (1)

Tricking a headless fruit fly into thinking that it is about to be swatted is quite impressive. Unfortunately, the neural response to the light was slow, and the process required a multicomponent system, (2) hence it never caught on. Miesenböck had shown that light could be used to activate genetically specified populations of neurons; now all that was required to make this one of the most important techniques of this era was a system that would be simpler and faster in its response to light than the optically gated system used by Miesenböck's group.

Many organisms, such as bacteria, algae, and plants, have molecular switches that respond to light; spanning membranes, they control the passage of positively charged ions like calcium and sodium and negatively charged chloride ions in and out of the cells. If the genes of any of these light switches could be expressed in the cell membranes of a neuron and responded to light in an ultrafast manner, they could act as selective gatekeepers for ions like calcium that cause the release of neurotransmitters in the neuron.

For more than 150 years, it has been known that motile microalgae respond to light (figure 10.1). The desired system was found in these algae. Back in 1991, long before Miesenböck used rat rhodopsin to activate fruit fly neurons, Peter Hegemann, a biophysicist at the Humboldt University of Berlin, showed that these algae rapidly produced a small electric current when exposed to light and suggested that fast production of the photocurrent was a result of the light receptors and ion channels being

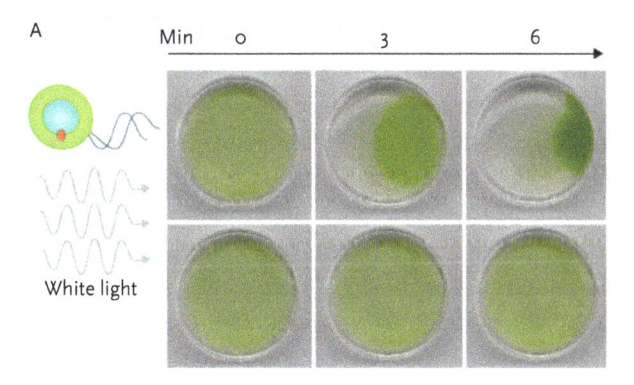

Figure 10.1 Wild-type motile microalgae, *Chlamydomonas reinhardtii*, respond to light and aggregate on right side of dish (top). Channelrhodopsin-defective mutants do not respond to light. Found in the light-sensitive "eye spot" of the algae, the function of channelrhodopsin is to allow calcium ions to enter the algae cells when they are exposed to blue light, thereby controlling their swimming and feeding behavior (bottom). (3)

intimately linked in a single protein complex. However, for 10 years the scarceness, instability, and intricacy of the proteins prevented Hegemann from purifying the protein complex. That changed when his group found the genes for two rhodopsins in the microalgae Chlamydomonas. In collaboration with Georg Nagel's group at the Max Planck Institute of Biophysics in Frankfurt, Germany, they expressed the genes in frog eggs and human kidney cells. In both cases the new proteins, named *channelrhodopsins*, responded very quickly to light by opening up ion channels and letting positively charged ions into the cell. Hegemann and Nagel realized that, due to their simplicity and rapid response, the channelrhodopsins would be very useful tools. These researchers remarked on their potential in a 2003 paper, and by 2006 many groups around the world were using channelrhodopsins for their light-gated cation conductance properties. According to Hegemann and Nagel, "The recent success of optogenetics is to a large part based on the simplicity of the merely 315 amino acid long channelrhodopsin2 fragment." (3, p. 175) There are many interesting early patents on the channelrhodopsin2 fragment (ChR2) and its uses in neurons. Hegemann, Bamberg, and Nagel have the most important ones: patents that involve ChR2 working as an on switch in a large variety of cells, in a large number of species.

One of the groups that began to work with channelrhodopsins was that of Karl Deisseroth. While doing the research portion of his MD/PhD in the lab of Richard Tsien, a brother of Roger Tsien, Deisseroth met Ed Boyden, who was a PhD student in the same lab. It was there, in 2000, that they first spoke about the possibility of devising a neural light-driven on/off switch. Nothing much came of their idea until 2004, when Deisseroth was doing postdoctoral research at Stanford University and Boyden was still doing his PhD work. When they read and discussed the Nagel and Hegemann paper describing ChR2 and its properties, they realized this might be the fast-responsive light switch they were looking for, and so Deisseroth contacted Georg Nagel, requesting the ChR2 gene. The Germans promptly sent a fusion gene of the channelrhodopsin with yellow fluorescent protein to Deisseroth at his postdoctoral position at Stanford. Because Boyden and Deisseroth were not yet independent researchers, they had to do all their optogenetics research after hours. To see whether channelrhodopsin was the molecular light switch they were looking for Deisseroth optimized the gene for viral insertion into hippocampal cell cultures. Figure 10.2 shows the resultant hippocampal neurons; the fluorescence confirms that the channelrhodopsin was expressed in these cells. (4) Ed Boyden, working in Tsien's lab using optical equipment that was part of his PhD research, was ready to test the light response of the modified hippocampal neurons.

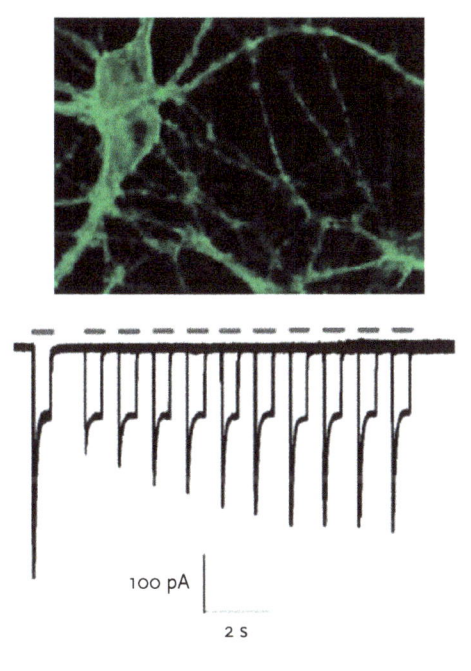

Figure 10.2 Hippocampal neurons expressing the channelrhodopsin–yellow fluorescent protein construct (top). Ten overlaid current traces recorded from a hippocampal neuron illuminated with pairs of 0.5-second light pulses (indicated by gray bars on top), separated by intervals varying from 1 to 10 seconds. Each peak represents a current produced in response to the 0.5-second light pulse (bottom). (4)

The very first glowing neuron he tested fired in response to light excitation. According to Boyden, "During that long, exciting first night of experimentation in 2004, I determined that ChR2 was safely expressed and physiologically functional in neurons. The neurons tolerated expression levels of the protein that were high enough to mediate strong neural depolarizations. Even with brief pulses of blue light, lasting just a few milliseconds, the magnitude of expressed-ChR2 photocurrents was large enough to mediate single action potentials in neurons, thus enabling temporally precise driving of spike trains. Serendipity had struck—the molecule was good enough in its wild-type form to be used in neurons right away. I e-mailed Karl, 'Tired, but excited.' He shot back, 'This is great!!!!!'" (5, p. 32)

In January 2005, Deisseroth was hired for an assistant professor position in the Department of Bioengineering and Psychiatry at Stanford University. In March of the same year, Boyden joined him as his postdoctoral researcher, and in April they submitted a paper on the work to *Science*, but it was rejected. Next they submitted their results to *Nature Neuroscience*, and in July the paper, entitled "Millisecond-Timescale, Genetically Targeted Optical Control of Neural Activity," was accepted. (4)

Interestingly, although Ed Boyden completed his PhD only 2 months after the paper was published, and at least half the experiments were done in Richard Tsien's lab, this was not acknowledged in the paper, and Stanford is listed as Boyden's place of work. The *Nature Neuroscience* paper was the first of a number of papers that appeared nearly simultaneously; the others described research in which channelrhodopsin was inserted into retina, chick spinal cord, and mammalian cells. Nagel and Hegemann had found a fantastic genetic light-controlled on switch for neurons. All these papers showed that light can be used to activate channelrhodopsin in neurons without the addition of any supplemental chemicals. A new field was born, and like all newborns it needed a name; Deisseroth came up with *optogenetics*, and it has stuck.

Thanks to channelrhodopsin, it was now possible to excite single neurons selectively and repeatedly. However, it would take 2 years before an off switch was found. It was found simultaneously by Deisseroth at Stanford and Boyden, who had moved to MIT, where he now had an academic position. Both used the gene for halorhodopsin, a protein found in very primitive bacteria, *Natronomonas pharaonis*, that thrive in salt flats in the Sahara Desert, where the protein allows the chloride ions to enter the bacterium upon exposure to yellow light. (6, 7) Negatively charged chloride ions neutralize positive calcium ions. The neurons now had an on switch (blue light → positive ions → action potential) and an off switch (yellow light → negative ions to neutralize the positive ions).

Deisseroth and Boyden both chose to demonstrate the utility of their off switches by genetically modifying *C. elegans*, a worm often used in scientific studies. *C. elegans* has 302 neurons whose functions are well known, which makes it an excellent model system for neural studies. Deisseroth and his co-workers chose to modify a neuron responsible for movement. He placed his genetically modified *C. elegans* in a water-filled petri dish. In order to survive, the roundworm has to swim; it is not in its best interest to rest. The only way to stop *C. elegans* from swimming would be to switch off the neurons responsible for movement. It worked: no light and the worm wiggled; yellow light and it stopped (figure 10.3). (7) In case you are concerned about the worm's welfare during these experiments, fear not; *C. elegans* is tough. In February 2003, scientists at the Kennedy Space Center discovered that some of the worms on the space shuttle *Columbia* survived its explosion. Messing around with its ability to wiggle about in water is not going to bother this little worm much. Later, after both papers had been published, the two groups teamed up to file a patent on the use of halorhodopsin as a light-driven off switch in neurons. (8)

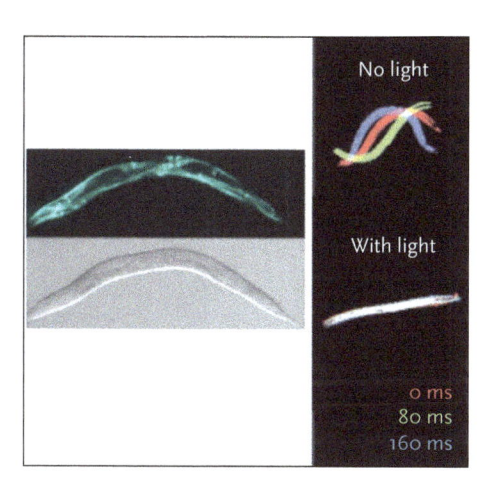

Figure 10.3 Top left: Fluorescence of halorhodopsin tagged with cyan fluorescent protein in the body wall muscles of *C. elegans*. When halorhodopsin is photoactivated with yellow light, chloride ions are released and the muscle neurons no longer fire; consequently the swimming behavior of *C. elegans* in liquid media is inhibited. Right panel: Animal postures from three consecutive film frames with and without yellow light were superimposed. With no light, the worm wiggles around, swimming for its life; while with yellow light, its neurons are inactivated by the chloride pump, and it is completely still. (7)

Now, no self-respecting neuroscientist, especially one who practices psychiatry and sees patients once a week, is happy with controlling the negligible mind of a worm. Thus it is not surprising that Deisseroth has progressed from worms to mammals and that by 2008 he and his group were using fiber optics to send mind-controlling blue light to the brains of rats, thereby manipulating their desire to run in circles. Figure 10.4 shows the six steps used to control the neurons in the brain of a live rat. A genetic construct is created that is composed of the "on switch," which is normally ChR2 and a promoter that will ensure that the ChR2 is expressed only in the neurons of interest (step 1). This construct is inserted into a benign virus (step 2) that is injected into the brain of the rat (step 3). The virus widely infects the brain, but the ChR2 is expressed only in the subgroup of neurons that have the necessary machinery to turn on the promoter in the original construct. A transgenic rat created in this way is ready for some optogenetics experiments; however, the neurons that need to be activated are in the brain, protected by the skull and its fur. No blue light is going to penetrate the fur and bone, and so in the fourth step a fiber-optic cable is threaded through the skull and glued in place with tooth cement. Laser light with the correct frequency to open the channel protein expressed in the membrane of the neuron is passed down though the optic fiber, so that the open channel allows for an influx of calcium or sodium ions and the neuron fires (step 5). The electrophysiological consequences of the action potential caused by the light can be measured

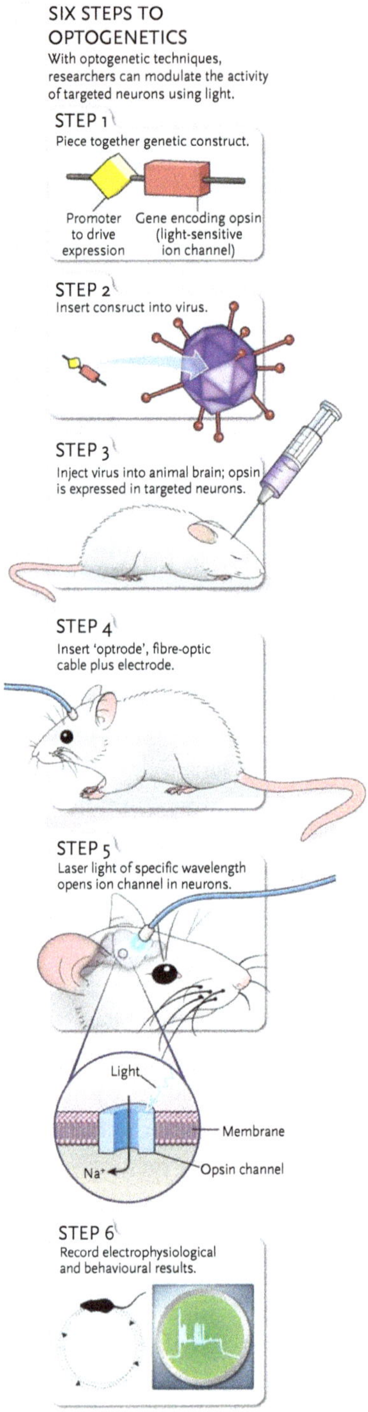

Figure 10.4 Illustration of the six steps of optogenetics described in the text. (9)

with an electrode that was inserted into the skull with the optic fiber, and video cameras linked to computers can be used to monitor the behavioral results, in this case running in circles (step 6). (9) In an article in *Nature Protocols* from 2010, Deisseroth's group gives a much more detailed and realistic description for the optogenetic integration of neural circuits. The procedure is broken down into 70 steps that can be completed in 4 to 5 weeks. (10)

In a paper in *Nature* in January 2008, Karel Svoboda, whom we met in chapter 9, and his colleagues reported how they have used optogenetic light pulses to train mice to get a reward from a specific port. The aim of the experiments was not to train the mice but to show the minimum number of neurons required for learned behavior. (11) The Svoboda, Deisseroth, and other labs working in the field regularly release videos of their experiments, typically added as supplementary material to research papers. It is surprising that none of these videos have gone viral. I remember getting chills the first time I saw the video of the mouse described in the prior paragraph. The video begins with a brown mouse grooming itself in a large, white plastic container. The mouse seems to have a sizable pimple on its head. Within seconds, however, it became obvious that this is not a pimple because it lights up with an eerie blue glow. It's the optic fiber sending blue light to the genetically altered neurons in the motor cortex. As soon as the little nub on its head lights up, the mouse stops grooming itself and starts running around the container in a counterclockwise direction. When the light is turned off, the mouse stops and resumes its grooming behavior. I find it amusing that the mouse in the video is a standard brown lab mouse, but the one depicted in most media articles was the one shown in figure 10.5, a rat with shiny snow-white fur.

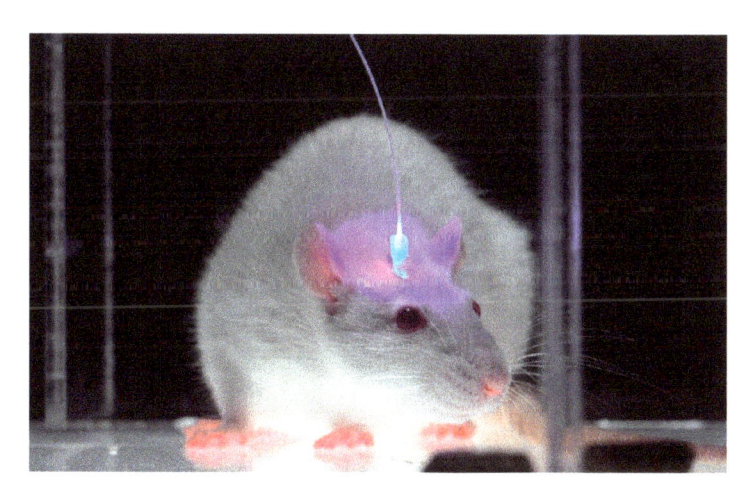

Figure 10.5 White rat linked to optic fiber that can transmit blue light to brain cells containing the algae protein channelrhodopsin. The blue light opens the ion channels in the channelrhodopsin, producing current and triggering a neural response.

Although most past and present optogenetics experiments are designed to understand the brain and how it functions, here I will present some examples related to Parkinson's disease, anxiety, blindness, and heartbeats; these are examples that show how optogenetics may one day be used to control diseases.

Starting in the 1990s, a treatment called deep *brain stimulation* was used to control tremors and stiffness in patients with severe forms of Parkinson's disease. It is a rather extreme solution in which a device similar to a pacemaker is implanted deep in the brain, in a tiny region called the *subthalamic nucleus*. There it stimulates the brain with carefully regulated electric pulses. Although the technique is quite effective, there are some drawbacks, particularly since it stimulates all surrounding neurons indiscriminately, and the medical basis of the treatment is unknown. (*12*) To get a better understanding of the underlying medical basis of deep brain stimulation, Deisseroth and his colleagues used viral gene transfer techniques to create a large number of mice, all having different neurons in their subthalamic nucleus modified with light-sensitive transmembrane ion channels. No matter which neurons they modified, the blue light transmitted through the fiber-optic cord had no effect. Out of desperation, they investigated axons leading to the subthalamic nucleus from the outer edges of the brain and hit the jackpot. Stimulating the connections leading to the subthalamic nucleus with high-frequency pulses cured the mice of their parkinsonian symptoms, and the therapeutic effects were immediate and completely reversible. Interestingly, low-frequency light stimulation had the opposite effect, aggravating the parkinsonian symptoms. (*13*) Using a microelectrode carefully implanted in the subthalamic nucleus, the researchers were able to detect neural activation deep within the brain upon exciting neurons located at the surface of the brain that were connected to the subthalamic nucleus with blue light (figure 10.6). According to Deisseroth, "Pointing to these axons that converge on the region opens the door to targeting the source of those axons. This insight leads to deeper understanding of the circuit and could even lead to new kinds of treatments. Because these axons are coming from areas closer to the brain's surface, new treatments could perhaps be less invasive than deep-brain stimulation." (*14*)

Most of the mouse models described in this book were obtained from the Jackson Laboratory, which supplies Brainbow, GCaMP, channelrhodopsin, and GFP (419 different varieties) modified mice. You can get mice that exhibit symptoms of Parkinson's disease, such as those used in the experiments described previously, and you can buy anxious mice. These nervous mice are slower to explore the mysteries of a new space, spend

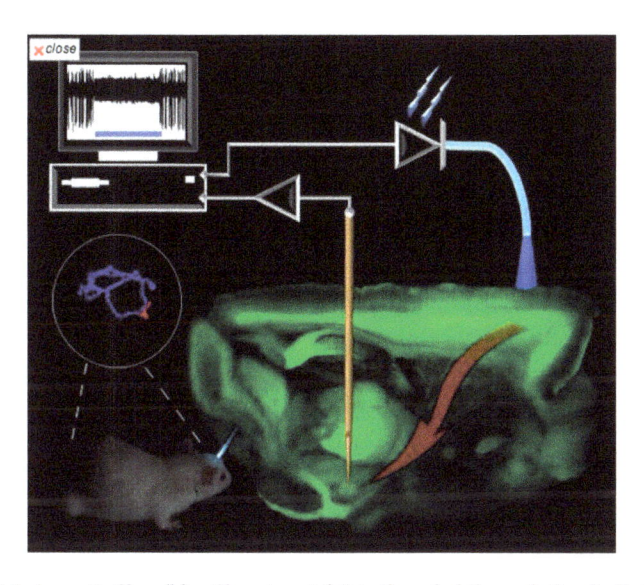

Figure 10.6 An optic fiber (blue fiber, top right) is threaded through the skull of a mouse so that high- and low-frequency flashes can be delivered to the ends of neurons that lead from the surface of the brain to the subthalamic nucleus of a mouse with severe Parkinson's disease. At the same time using a strategically implanted microelectrode (yellow electrode), the researchers can detect activation of cells deep within the subthalamic nucleus of the brain. (*Science Daily*, March 19, 2009.)

less time in bright open spaces, and choose to isolate themselves in an empty cage rather than join other mice in neighboring cages. Many studies, including fMRI studies, have shown that the amygdala is the part of the brain associated with fear. Of course, anxiety is not limited to mice that have been bred to be fearful; at sometime in their lives, a quarter of all people will experience sufficient anxiety for it to be classified as a diagnosable psychiatric disorder, making anxiety the most common psychiatric disorder. Anxiety disorders often can lead to depression and alcohol dependence. Deisseroth and his group (once again) were able to use optogenetics to change anxious mice into brave, bold, adventurous mice or into even more fearful ones. Deep within the amygdala, which is associated with fear, they found some neural connections that were responsible for a reversible antianxiety effect, making them perfect targets for optogenetics and modification with channelrhodopsin and halorhodopsin. Illuminating these modified connections with blue light activated the channelrhodopsin, causing the antianxiety neurons to fire and producing a mouse that overcame its fears and explored open areas. Changing the light from blue to yellow completely changed the character of the mouse, which just hunkered down in a dark corner giving in to its anxieties. When the blue light was applied more broadly, activating more neurons in the amygdala, the antianxiety effect was erased. A similar effect could be reducing the efficiency of current pharmaceuticals, which are much less specific. In an article in the *New York Times*, David Anderson, a biology professor at the California Institute of Technology, compared the drugs' effects to a sloppy oil change: "If you dump a gallon of oil over your car's engine, some of it will dribble into the right place, but a lot of it will end up doing more harm than good." (*15*, p. D1) The amygdala is structured similarly in humans and mice; perhaps there is hope for those of us suffering from anxiety because our children have overactive antianxiety genes that turn them into daredevils.

Many may argue that controlling the heart is more important than controlling the mind. Optogenetics is up to that task too, since it can be used to activate any electrically excitable cell, including smooth muscle. Given the wide distribution and cost associated with heart disease, it is not surprising that many research groups around the world are applying optogenetic techniques to heart-related problems. It is well known that cardiomyocytes are the heart cells that regulate the heartbeat, and they have therefore become the targets for blue light control. In 2010, Didier Stainier and his colleagues at the Cardiovascular Research Institute at the University of San Francisco reported that they have used light, channelrhodopsin, and halorhodopsin to act as a pacemaker in zebrafish cardiomyocytes. (*16*) Although the Deisseroth group's focus is mainly on neurological diseases, in September 2011 it released the blueprints of the first light-regulated human pacemaker. (*17*) Oscar Abilez, a postdoctoral researcher in the Deisseroth group, inserted the gene for ChR2 into human embryonic stems cells and then transformed them into cardiomyocyctes. In a healthy human heart, the pacemaking cells are located on the top of the heart, and the contractions they control radiate down to and around the heart. In creating an alternative transgenic pacemaker, the newly created cardiomyocyctes were not randomly injected into the heart; instead, a computer was used to model the cardiomyocyte-induced heartbeat and to find the best location to inject the light sensitive cells. The researchers were looking for, and found, a location that would result in contractions identical to those observed in the healthy human heart. The ultimate aim of this research is to find a less invasive pacemaker to replace the millions of pacemakers that regulate faulty hearts all around the world. "We might, for instance, create a pacemaker that isn't in physical contact with the heart," said coauthor Christopher Zarins, professor emeritus of surgery and director of the Stanford lab where Abilez performed the experiments. "Instead of surgically implanting a device that has electrodes poking into the heart, we would inject these engineered light-sensitive cells into the faulty heart and pace them remotely with light, possibly even from outside of the heart." "And because the new heart cells are created from the host's own stem cells, they would be a perfect genetic match," added Abilez. "In principle, tissue rejection wouldn't be an issue." (18)

Ed Boyden and Alan Horsager, a neuroscientist at the Keck School of Medicine at the University of Southern California and MIT, have started a new biotech company, EOS Neuroscience, to help take optogenetics from the lab to the clinic. "It's really exciting to think of the clinical applications opened up by the ability to control neurons by light," says Boyden. So far the company has no marketable products, but it has some preliminary results that look very promising. It started looking for commercial applications in diseases related to the eye; according to Boyden, "The eye, which can access light from the outside world, is a perfect test bed for the use of optogenetic tools for treating intractable disorders." In April 2011, the EOS scientists reported that they used optogenetics to restore the ability of blind mice to differentiate between light and dark. The blind mice had retinitis pigmentosa, a disease in which light-sensitive cells in the retina are destroyed and the brain no longer receives visual image information. Alan Horsager and his neuroscientists injected a virus with the channelrhodopsin gene into light-insensitive cells on the surface of the retina. Thanks to the rhodopsin, the modified cells could sense light and transmit the information to the appropriate section of the brain. To test their sight, the mice were placed in a water maze that had a light illuminating its exit. The blind mice swam around randomly and took a long time to locate the exit, while the sighted mice and the genetically modified mice were able to see the light and head straight for the exit. Besides helping blind mice navigate water mazes (and perhaps saving them from the farmer's wife who might cut off their tails with a carving knife), this research has important human implications given that more than 100,000 people in the United States have lost their sight to retinitis pigmentosa. (19)

Like fluorescent protein technology, optogenetics is used primarily to increase our understanding of biochemical processes occurring in live organisms; however, it can do more. The retinitis pigmentosa, Parkinson's, and skittish mouse examples illustrate that the joint use of light to image and control neural circuits with optogenetics has taken us from the point where we can illuminate disease to a point where we can realistically hope to cure neural disorders. Just as fluorescent proteins have grown from simple protein tracers to in vivo calcium monitors and indicators of protein misfolding, finally giving birth to new fields such as optogenetics, so optogentics is maturing and spawning new applications, including one developed by Feng Zhang, who while at MIT devised a technique that uses a system of light-sensitive proteins to switch on and off the expression of specifically targeted genes.

The development of fluorescent proteins and their subsequent use in all areas of science are an excellent examples of basic science leading to practical biotechnological and medical applications. More than 40 years ago, a small band of researchers were interested in elucidating the photochemistry responsible for the green light emitted by the crystal jellyfish. Together they caught more than a million jellyfish and spent more than three-quarters of a century studying jellyfish bioluminescence. Now green fluorescent protein is the microscope of the twenty-first century. In technicolor, it allows us to see things we have never been able to see, thereby completely changing the way we approach science and medicine. Thanks to GFP, researchers can follow the spread of cancer cells in real time in live mice, and genetically modified mosquitoes with glowing gonads may curb the spread of malaria. The jellyfish protein has also led to new techniques, such as optogenetics, that not only grew from the GFP revolution but also use the fluorescent protein as a marker.

I like to think of research projects as puzzles. In mysterious ways, nature has created exquisite multidimensional puzzles for us to solve. But nature has been cruel, too; the pieces don't come in a box, with the number of pieces listed and an image of the completed puzzle on the lid. To solve any of nature's puzzles, researchers need to find the pieces before trying to place them in the puzzle.

Fluorescent proteins are ubiquitous in scientific and medical research because they have allowed researchers to find new puzzle pieces that otherwise would have remained hidden. For example, they have been used to show how stem cells migrate from a mouse embryo to a distressed heart and can light up synapses in the brain. But they can also be used to light up more esoteric pieces of the puzzle. In fact, the examples given in this book are the exception rather than the rule; they were chosen because the fluorescent proteins highlighted a very interesting part of a disease-related puzzle that was easy to understand and explain. Although it is impossible to page through an issue of *Science* or *Nature* without seeing an image showing glowing fluorescent proteins, there are many more uses of GFP that don't make it into these magazines and don't even warrant a color image in a publication. In real life, fluorescent proteins have become just another tool in the lab that are routinely used and are mentioned only in the methods section of research papers, often without any specific citation. This makes it difficult to establish exactly how often they are used.

The most important lessons we can learn from channelrhodopsins and their older siblings, the fluorescent proteins, do not come from the flashy applications that grab the public's attention and go viral, the examples from this book, or the everyday uses that light up labs all around the world. Instead, these wondrous proteins remind us of the importance of fundamental research and the necessity to protect nature's diversity, for we have no way of knowing which endangered species contain the keys for unlocking science's big secrets. These sentiments are expressed in a quotation from Karl Deisseroth that strongly echoes the portion of Roger Tsien's Nobel speech that concludes chapter 1:

> The lesson of optogenetics is that the old, the fragile and the rare—even cells from pond scum or from harsh Saharan salt lakes—can be crucial to comprehension of ourselves and our modern world. The story behind this technology underscores the value of protecting rare environmental niches and the importance of supporting true basic science. We should never forget that we do not know where the long march of science is taking us or what will be needed to illuminate our path.
>
> Karl Deisseroth, "Controlling the Brain with Light," *Scientific American*, October 20, 2010

REFERENCES

1. Miesenbök, G. (2010). Re-engineering the brain. TED Global, Oxford, July.
2. Lima, S. Q., and Miesenböck, G. (2005). Remote control of behavior through genetically targeted photostimulation of neurons. *Cell 121*, 141–152.
3. Hegemann, P., and Nagel, G. (2013). From channelrhodopsins to optogenetics. *EMBO Molecular Medicine 5*, 173–176.
4. Boyden, E. S., Zhang, F., Bamberg, E., Nagel, G., and Deisseroth, K. (2005). Millisecond-timescale, genetically targeted optical control of neural activity. *Nature Neuroscience 8*, 1263–1268.
5. Boyden, E. S. (2011). The birth of optogenetics: An account of the path to realizing tools for controlling brain circuits with light. *The Scientist 25*, 30–36
6. Han, X., and Boyden, E. S. (2007). Multiple-color optical activation, silencing, and desynchronization of neural activity, with single-spike temporal resolution. *PLOS One 2*(3), e299.
7. Zhang, F., Wang, L. P., Brauner, M., Liewald, J. F., Kay, K., Watzke, N., Wood, P. G., Bamberg, E., Nagel, G., Gottschalk, A., and Deisseroth, K. (2007). Multimodal fast optical interrogation of neural circuitry. *Nature 446*, 633–634.
8. Boyden, E. S. (2011). A history of optogenetics: The development of tools for controlling brain circuits with light. *F1000 Biology Reports 11.*, B3–11.
9. Buchen, L. (2010). Neuroscience: Illuminating the brain. *Nature 465*, 26–28.
10. Zhang, F., Gradinaru, V., Adamantidis, A. R., Durand, R., Airan, R. D., de Lecea, L., and Deisseroth, K. (2010). Optogenetic interrogation of neural circuits: Technology for probing mammalian brain structures. *Nature Protocols 5*, 439–456.

11. Huber, D., Petreanu, L., Ghitani, N., Ranade, S., Hromadka, T., Mainen, Z., and Svoboda, K. (2008). Sparse optical microstimulation in barrel cortex drives learned behaviour in freely moving mice. *Nature 451*, 61–64
12. Medtronic. (2013). Shaking before/after DBS for Parkinson's treatment. Video. www.youtube.com/watch?v=a_4_DvquSYQ.
13. Gradinaru, V., Mogri, M., Thompson, K. R., Henderson, J. M., and Deisseroth, K. (2009). Optical deconstruction of parkinsonian neural circuitry. *Science 324*, 354–359.
14. Orenstein, D. (2009). Stanford Study improves insights into Parkinson's disease and possible treatments. *Stanford School of Medicine News Release*, http://med.stanford.edu/news_releases/2009/march/deisseroth.html.
15. Schoonover, C. E., and Rabinowitz, A. (2011). Control desk for the neural switchboard. *New York Times*, May 16, D1.
16. Arrenberg, A. B., Stainier, D. Y. R., Baier, H., and Huisken, J. (2010). Optogenetic control of cardiac function. *Science 330*, 971–974.
17. Abilez, O. J., Wong, J., Prakash, R., Deisseroth, K., Zarins, C. K., and Kuhl, E. (2011). Multiscale computational models for optogenetic control of cardiac function. *Biophysical Journal 101*, 1326–1334.
18. Myers, A. (2011). Researchers create first human heart cells that can be paced with light. *Stanford—School of Medicine*. http://med.stanford.edu/ism/2011/september/cardiomyocyte.html.
19. MIT Tech TV. (2011). Blind mice, no longer. www.youtube.com/watch?v=jY5Aynh1-cU video.

INDEX

Adelman, Zach 96–97
Aedes aegypti 77, 90–91, 94–104
Aequorea victoria 1–3, 6, 49
   aequorin 6–8
Allison, Anthony 69
Alphey, Luke 91, 101–102, 105
ALS 186, 188–190
Alzheimers 181–183, 190–191, 193
amyloid plaques 181–183
angiogenesis 119–121
Anopheles gambiae 56, 59, 65, 74–75, 77, 80, 82
Anticancer Inc 15, 20, 116–121, 130–132
antimalarial drugs
   Artemisinin 54, 70–71
   Decoquinate 64
   Methylene Blue 72
   Quinine 55, 69–70, 84
   Sanaria vaccine 73
   transmission-blocking vaccines 72
anxiety, optogenetics 212–215
Aravinthan, Samuel 196–197
astroglia 179–180
ATP 16, 29–30

bacterial cancer therapy 131
Barré-Sinoussi, Françoise 156–157
BioArt competition 179
Bird Flu, H5N1 138, 140, 143–150
Blackshaw, Seth 175–176
BOLD 189–190
Boucher, Richard 139
Boyden, Edward 204, 206, 215
Brainbow 170–172, 176, 190, 212
Brocolli, Vania 186, 188

C. elegans
   first GFP use in 9–10, 13, 15
   heart, ultradian rhythms 24–25
   misfolded proteins 185
   electric field 196–197
   optogenetics 208–209
calcium sensors 7
   Cameleon 191–192, 194, 197
   GCaMP 203, 212
   GCaMP3 194–195
cancer
   chimney sweeps 111
   dinosaur, cancer 110
   stem cells 124–126
   T-cells 114–115
   word origin 110
cardiomyocytes 29–35, 214
CD4+ T cells 158–162
Chagas, Carlos Justiniano Ribeiro 46–47
Chalfie, Martin 7, 9–14, 17–19, 116, 198
Channelrhodopsin 205–208, 211–216
Chaudhry, Hina 32–34
Chen, Benny 40
Cheng, Heping 29
CLARITY 168, 172
Contag, Christopher 113, 116
cotton pest moth 83–84
Cormier, Milton 7–8
Cruzigard 47
Crisanti, Andrea 78–79, 82
Crystal Jellyfish. See *Aequorea victoria*
Cypridina 5–6

DDT 74–75, 84, 89, 94, 98
DEET 77

Deisseroth, Karl 204, 206–209, 211–213, 217
dengue hemorrhagic fever 92–93, 98, 142
dopaminergic neurons 186, 188
Doyle, Michael 95, 104–105
DsRed 49–50, 82, 84, 132
Duffy antigen 67
Durvasula, Ravi 47–50

early-onset dystonia 185–186
EGFP 14–15
Elliott, David 35
Euskirchen, Ghia 11–12

feline immunodeficiency virus 163
Fidock, David 72
Florida Keys Mosquito Control District 95, 104
fluorescence-guided surgery 130
fluorescence resonance energy transfer 192
fluorescent influenza virus 137, 140
fMRI 189–190, 191, 198, 213
Fouchier, Ron 144–145
fowlpox 146–147
Frischknecht, Friedrich 65
fruit fly
    Alzheimer's disease 182
    carbon dioxide detection 78
    GFP first used in 13
    optogenetics 204–205
    sterile insect technique 80, 81
    Wolbachia 99
Fucci, cell cycle indicator 122

Gallo, Robert 156–157
Garcia-Sastre, Adolfo 137
Giancotti, Filippo 127
glaucoma 179–180
glial cells 125, 179
Golgi, Camillo 168–170
Greenwalt, Dale 57
Griffiths, Elinor 30
Guevara, Palmira 51

Hahn, Beatrice 58–59, 155
Halorhodopsin 208, 213
Hazelrigg, Tulle 13–14

heart
    attack 29–31, 33–34, 36, 47
    beat 27–29
    energy 29–30
    transplants 36–40, 51
Hecht, Michael 181–183
Hegemann, Peter 204–206, 208
hemagglutinin 138, 140–141
hemoglobin 57, 60, 62
    sickle cell anemia 68–69
HER2 126
Herceptin 126
Hill, John 111
history
    cancer 110–112
    Chagas 46–47
    dengue fever 92–93
    HIV/AIDS 155–157
    influenza 141–144
    malaria 54–56
Hoffman, Robert 15, 20, 116–119, 130–131
HTLV-3 157

Ikuta, Kazuyoshi 140–141

Jacobs-Lorena, Marcelo 66, 67
Johnson, Sterling 190

Kaede 31–32
Kawakami, Yasuhiko 31–32
Kawaoka, Yoshihiro 144–145
Keller, Phillipp 197–198
Ken, Suzuki 36
Kim, Jinhyun 176–177
kissing bug 43–44, 45–48
Knols, Bart 77
Kotlikoff, Michael 27–29

LAV 156–157
Laveran, Charles 55–56
Lewy bodies 184
light-sheet microscopy 198
Looger, Loren 193–195
Lou Gehrig's Disease 181, 186
luciferase 15, 16, 30, 63, 64, 72, 94
    cancer 113–116, 128
    salmonella 113, 131–132
    Xenogen 113
Lukyanov, Sergey 49–50

Massagué, Joan 114–115
Matz, Mikhail 49
Mempel, Thorsten 161
Ménard, Robert 62
Metformin 183
mGrasp 176–179, 190
Miesenböck, Gero 203–205
Miller, Freda 183
mitochondrial communication 30
mosquito fossil 57
mucus—flu 139
myoblasts 36
Miyawaki, Atsushi 31, 122, 171–172, 191–192
Montagnier, Luc 156–157
Mota, Maria 63

Nabel, Gary 159–160
Nagel, Georg 204, 206, 208
nanotubes 160–161
Nedergaard, Maiken 180
neuraminidase 138, 140
Nobel Prize
 chemistry 2008 7, 17–20, 157, 191
 medicine 1902, Ross 55
 medicine 1906, Cajal, Golgi 169
 medicine 1907, Laveran 56
 medicine 1948, Müller 74
 medicine 2008, Montagnier 157
 medicine 2008, Barré-Sinoussi 157

O'Neill, Scott 89–90, 99–101
Oxitec 83–84, 90–91, 101–106

P. berghei 62–67
P. falciparum 54–59, 62, 65–68, 70–74, 80, 155
P. vivax 54, 61, 63, 65–67
pacemaker, optogenetics 214
Parada, Luis 125
Parkinson's 40, 126, 181, 183–186, 193
 optogenetics 212–213, 215
photoactivatable GFP 29–30
Plasmodium 54–56, 58–67, 69, 71–73, 75–76, 79–80
 fluorescent 60, 62–64, 66
 life cycle 59–63
 luciferase 63, 94
Poeschla, Eric 163
Pott, Percival 111

Prasher, Douglas 7–12, 14, 16–20, 157, 177
Prather, Randy 37–39
Prudêncio, Miguel 60, 63–64
Prugnolle, Franck 59

Rabinovich, Brian 114
real time *in vivo* imaging 119
retinitis pigmentosa 215
Riehle, Michael 66, 95
Riley, Paul 34
Ross, Ronald 55–56

Santiago Ramón y Cajal 166–171, 176, 190, 198
 Vacation Stories 198
Sang, Helen 147–150
Scale2 171–173
screwworm 80–81
selfish jumping genes 78
Simian immunodeficieny virus 37, 155
Skinner, Michael 146
Spanish flu, 1918 flu 141–143, 150–151
spinal motor neurons 186–187
Stainier, Didier 214
stem cells
 cancer 124–127
 heart muscle cells 31, 33–36
 neurons 183, 186–188
St. Leger, Raymond 79–80
sterile insect technique 80–83, 91, 101–102, 106
superfolder 181
swine flu, H1N1 138–140, 143, 150–151
Svoboda, Karel 174, 176, 193–195, 211

T. cruzi 43–44, 46–48, 50–51
T. rangeli 50–51
Talbor, Robert 70
Tank, David 195–196
tanycytes 175–176
Thioridazine 126
Tiley, Laurence 147–150
Torsin A 185–186
Tsien, Richard 191, 197, 206, 208, 217
Tsien, Roger 7, 14–15, 17–20, 157
Turner, Stephanie 77–78

ultradian rhythms 23–26

Verdin, Eric 160
Vetter, Monica 179
Villu Maricq, Andres 24–26
virus
   dengue 91–94, 96–101, 106, 157
   flaviviruses 92

Waldo, Geoffrey 181
Wichterle, Hynek 187–188
Wolbachia 99–101, 106
working memory 195–196

zebrafish 32–33, 197, 214

Ingram Content Group UK Ltd.
Milton Keynes UK
UKHW020750250723
425740UK00002B/41